Statistical Approaches for Landslide Susceptibility Assessment and Prediction

Sujit Mandal • Subrata Mondal

Statistical Approaches for Landslide Susceptibility Assessment and Prediction

Springer

Sujit Mandal
Department of Geography
Diamond Harbour Women's University
Sarisha, West Bengal, India

Subrata Mondal
University of Gour Banga
Mokdumpur, West Bengal, India

ISBN 978-3-030-06739-7 ISBN 978-3-319-93897-4 (eBook)
https://doi.org/10.1007/978-3-319-93897-4

This Springer imprint is published by the registered company Springer Nature Switzerland AG
The registered company address is: Gewerbestrasse 11, 6330 Cham, Switzerland

Preface

Landslide is one of the most destructive natural hazards which causes great damage to lives and properties all over the world. The destructive character of landslide phenomena has been observed in the Himalayan region, especially in Darjeeling Himalaya. The significant growth of human population and continuous expansion of communication lines and human settlement in Darjeeling Himalaya have been posing a tremendous pressure over mountain slopes, and have caused landslide phenomena for a long time. Researchers have paid much more attention to the destruction caused by landslides, and they have started to find out the probable causes of landslides in Darjeeling Himalaya and probable mitigation measures. In the present day, the effective measures with the application of scientific tools and techniques have become much more important in landslide studies. The destructive character of landslide phenomena in Darjeeling Himalaya encouraged me to prepare this book on *Statistical Approaches for Landslide Susceptibility Assessment and Prediction*. Several approaches and methods with regard to landslide studies and preparation of landslide susceptibility, hazard, vulnerability, and risk zonation maps have been developed worldwide. The identification of landslide locations and potential landslide susceptibility zones with the help of suitable data layers on GIS environment has been accepted by planners, developers, and researchers. The knowledge and experience on the causes and consequences of landslides and finding out the factors responsible for landsliding played a significant role in the study of landslide susceptibility, hazard, and risk. Earlier there were no important statistical approaches in landslide susceptibility assessment and prediction, and the potential landslide susceptibility zones of a rugged terrain were not possible to find out. But with the introduction of remote sensing techniques and GIS tools, it has become possible to prepare landslide hazard and susceptibility zonation maps for any geographical region.

Landslides are the downslope movement of rock, soil, and debris under the influence of gravity. The movement of earth materials downslope depends on geology, geomorphology, and hydrology of the concerned geomorphic unit. Landslide may be caused due to sudden occurrence of heavy showers or caused as a result of earthquakes, and is the outcome of several factors. On the basis of the materials present, landslide can be classified into earthslide, debris slide, rockslide, and mudslide. The movement of the materials occurs in different directions, i.e. diagonal, vertical, and lateral. But fall, slide, topple, flow, and spread are the common types of landslides. All these types of landslides cause great damage to human lives and properties and retard growth and development of a society. Considering the destructive impact of landslide phenomena, planners, policymakers, and researchers have decided to develop suitable approaches and methodologies to mitigate such hazardous events. In planning and development, the identification of suitable areas is very significant. To promote planning processes for mountain environments and landslide prone areas, the preparation of a landslide susceptibility zonation map is essential. Various statistical approaches with the help of RS and GIS are being applied to assess landslide susceptibility worldwide. To assess and predict landslide susceptibility of Darjeeling Himalaya, various statistical models such as frequency ratio (FR) model, modified information value (MIV) model, logistic regression (LR) model, artificial neural network (ANN) model, weighted overlay analysis (WOA) model, certainty factor (CF) model, analytical hierarchy process (AHP) model and fuzzy logic (FL) approach were used in conjunction with RS and GIS, GPS, satellite images, google earth images, topographical maps, historical landslide data, and rainfall data.

Chapter 1 of this book is about the major objectives, a detailed description of the study area of Darjeeling Himalaya, methodologies applied in the present study, existing literature corresponding to the objectives and themes of each chapter, data sources, a brief outline on the various statistical methods and its application in landslide susceptibility assessment and prediction, statement of the problem, and limitations of the study. Chapter 2 includes the preparation of various thematic data layers corresponding to the landslide conditioning factors of Darjeeling Himalaya. Various data layers, i.e. elevation, aspect, slope, curvature, geology, soil, lineament density, distance to lineament, drainage density, distance to drainage, stream power index (SPI), topographic wetted index (TWI), rainfall, normalized differential vegetation index (NDVI) and land use and land cover (LULC), are described in detail. To prepare all the data layers, SOI topographical maps, Google Earth image, satellite images (Landsat TM), geological map, soil map, and distribution of rainfall were collected from various sources and used properly on GIS environment. To assess the probability of landslide susceptibility of each class of all the landslide conditioning factors, frequency ratio was estimated with the help of the total number of pixels and the total number of landslide-affected pixels existing in each class. The estimation of frequency ratio value helped to perform frequency ratio model, information value model, index overlay model, and other statistical models and the preparation of landslide susceptibility zonation maps. Chapter 3 deals with the development of landslide susceptibility zonation maps using frequency ratio (FR) model and modified information value (MIV) model. Chapter 4 covers the performance of logistic regression (LR) model and identification of landslide susceptibility zones in Darjeeling Himalaya. To assess the probability of landslide susceptibility, logistic regression coefficient values were estimated for each class of all the landslide conditioning factors, and based on the said values, a landslide susceptibility zonation map was made on GIS environment. Chapter 5 represents the application of artificial neural network (ANN) model and landslide susceptibility study with the help of RS and GIS. The back-propagation training algorithm is the most frequently used neural network method and is the method used in this study. The back-propagation training algorithm is trained using a set of examples of associated input and output values. There are two stages involved in using neural networks for multisource classification: the training stage, in which the internal weights are adjusted, and the classifying stage. Typically, the back-propagation algorithm trains the network until some targeted minimal error is achieved between the desired and actual output values of the network. Once the training is complete, the network is used as a feed-forward structure to produce a classification for the entire data as well as to prepare landslide susceptibility map of Darjeeling Himalaya. Chapter 6 presents the application of weighted overlay analysis (WOA) model, certainty factor (CF) model, analytical hierarchy process (AHP) model and the assessment of landslide susceptibility of Darjeeling Himalaya. To prepare the zonation map of Darjeeling Himalaya, weighted overlay analysis was performed on GIS platform, and finally, a relationship was established between the prepared susceptibility map and all the thematic data layers. Certainty factor was calculated for each data layer based on the landslide inventory and the landslide occurrence frequency. The certainty factor for each pixel is defined as the change in certainty that a proposition is true (prior probability of having landslide in the study area) when given the evidence (conditional probability of having a landslide given a certain class of a thematic layer) for each data layer. AHP is a decision-making and semi-quantitative value judgement approach which serves as the objectives of the decision-makers. This process is employed to support the decision on the instability rank of the factors by estimating prioritized factor rating value (PFRV). Chapter 7 reveals a soft computing technique, known as fuzzy logic (FL), which has been applied to map landslides. Considering the determination of weightings (fuzzy membership values), this technique is either an expert-based (semi-quantitative) approach or a data-driven (quantitative) approach or a combination of both. Finally, in Chap. 8, receiver operating characteristics (ROC) curve was used to make a comparison between frequency ratio (FR) model, modified information value (MIV) model, logistic regression (LR) model, artificial neural network (ANN) model, weighted overlay analysis (WOA) model, certainty factor (CF) model, analytical hierarchy process (AHP) model and fuzzy logic (FL) approach and to find out the suitable models in landslide susceptibility assessment and prediction of Darjeeling Himalaya. This book will contribute a lot for students, teachers, and researchers to find out various statistical approaches in landslide susceptibility assessment and prediction. We would like to express our sincere thanks to all of our well-wishers who always supported us to be involved with such academic works. Specially, we would like to express our sincere thanks to all research scholars of University of Gour Banga for their relentless support throughout. We extend our sincere gratitude to National Atlas and Thematic Mapping Organisation (NATMO), National Bureau of Soil Survey and Land Use Planning (NBSS & LUP), Geological Survey of India (GSI), Survey of India (SOI), and United States

Geological Survey (USGS) Earth Explorer for providing adequate data for continuing the work without any problems. It is difficult for us to fully express our gratitude to our families, parents, brothers, sisters, and relatives for their sacrifice, constant love, cordial affection, and inspiration throughout the work.

Sarisha, West Bengal, India — Sujit Mandal
Mokdumpur, West Bengal, India — Subrata Mondal
January 2018

Contents

Chapter 1
Concept on Landslides and Landslide Susceptibility

1.1 Introduction

Landslide accounted for about 9% of all the natural hazards worldwide (Mousavi et al. 2011). It also caused approximately 1000 people deaths per year worldwide and about four billion US property damages (Lee and Pradhan 2007). In India, 15% (0.49 million sq. km) of the land area is affected by landslides including the Himalayan mountain regions, the Meghalaya plateau, and the Western Ghats (NDMA 2009).

Darjeeling Himalaya exhibits a wide range of diversity in terms of physiographic and socio-economic aspects. The Himalayan mountain is the result of tectonic movement which compressed the Tethys geosynclinals sediments in different phases of mountain building processes. The mechanism of the mountain building processes caused various thrust planes, i.e. and deformed the entire lithology throughout the Himalaya mountain range. The rugged Himalayan terrain is attributed with heterogeneous lithological composition, geomorphic unit, and soil type. As the Himalayan terrain is experienced with frequent tectonic events, there exists numerous lineaments, fractures, thrust plane, bedding plane, foliation plane, active faults, etc. All these micro-level geological character at local or regional level have made the mountain slope more susceptible to landslides. Since 1950, the depletion of forest cover due to expansion of human settlement, establishment of the tea garden area, and development of communication lines all over the Darjeeling Himalaya degraded the mountain environment as well as invited environmental hazards. This Himalayan region receives huge amount of orographic rainfall each year during monsoon period (June, July, and August) which is one of the most important triggering factors for causing environmental hazards. Not only that Darjeeling Himalaya is attributed with very high to moderate seismic intensity zones where medium level of vibration may induce havoc landslips and finally may cause great damage. 2011 earthquake in Sikkim-Darjeeling Himalaya caused

S. Mandal, S. Mondal, *Statistical Approaches for Landslide Susceptibility Assessment and Prediction*, https://doi.org/10.1007/978-3-319-93897-4_1

destructive landslide events which destroyed human lives and properties a lot in major district towns of Mirrik, Kalimpong, and Darjeeling. In the present day, the proper understanding in connection to landslides conditioning factors required for mountain slope management. Landslide is commonly regarded as spatially discrete and temporally dynamic phenomena. Detailed understanding of the slope soil geotechnical properties can provide reliable information in connection to landslides. In this case, generally considered landslide influencing factors and applied statistical models could not introduce the actual field situation as compared to the geotechnical approaches and methods. Ayalew and Yamagishi (2005) opined that landslides are being controlled by various geo-environmental factors which are the complex mass movement processes. Various approaches and methods have been applied by researchers to analyse slope instability applying geo-physical and geostatistical methods. Van Western et al. (2006) divided the spatially modelling of landslides into two approaches. First approach is dealt with the mathematical methods covering deterministic and dynamic modelling which are closely related to the physical mechanism. The physical mechanism discusses the geo-technical properties of rocks and soils which have pivotal role in stabilizing as well as destabilizing the mountain slope. Second approach is associated with location of previous landslides and landslide conditioning factors. This approach combines various landslide conditioning factors and identifies potential landslide susceptible areas over the space by applying discriminant analysis, multi-variate statistics, likelihood ratio, information value method, linear regression, etc. It is also a major concern to all of us to assess the landslide temporally. The temporal probability of landslide can be assessed with the help of two important procedures such as estimation of threshold rainfall and frequency of past landslide occurrences. Guzzetti et al. (2005) successfully applied frequency analysis to analyse the spatial distribution of landslide susceptibility on the basis of past landslide occurrences number. This is a data-driven technique with regard to landslide probability. Landslide hazards can also be assessed by using the information on the changes frequency-area or volume (Guzzetti et al. 2005). Crovelli (2000) and Guzzetti et al. (2005) applied the poisson probability model to figure out the expected occurrences of landslide probability at a specific time interval. The statistical methods in landslide studies use mathematical models and processes which integrate various landslide influencing factors and landslide inventory and finally prepare landslide susceptibility zonation map of a particular area with reasonable accuracy using statistical, mathematical, and GIS software. In this chapter, various statistical methods are to be discussed for landslide assessment and prediction.

The quantitative and statistical assessment of various phenomena and its logical description and analysis are the prime concerns in present day research of all scientific disciplines. In geography, the assessment of several geomorphic attributes and their logical description is required to establish a scientific principle for planning and development. To achieve this goal, proper scientific methods and techniques are to be incorporated lucidly. Researchers faced problems in landslide studies in India because of the lack of information required. Air photos and topographical maps are restricted for all border and problematic zones in the country including the entire Himalayan mountain region and coastal region. As a result of which landslides inventory map are not readily available for landslide susceptibility assessment and prediction. But, India has its own high resolution satellite images, e.g. Resourcesat-1, Liss-IV, Cartosat-1, which helped a lot to prepare landslide inventory maps and landslide susceptibility maps in using GIS software with reasonable accuracy.

In the present study of spatial distribution of landslide hazards of Darjeeling Himalaya information and knowledge-driven methods and techniques such as frequency ratio model (FRM), information value model (IVM), logistic regression model (LRM), artificial neural network model (ANN), index overlay model (IOM), certainty factor (CF) model, analytical hierarchy process (AHP) model, and certainty factor approach (CFA) were being adopted with suitable logic and principle using RS and GIS. Finally, a comparison between models was also to be made for its implementation in planning and development of the mountain region. To perform all these models as well as to prepare landslide susceptibility zonation map of Darjeeling Himalaya, various data layers corresponding to geology, geomorphology, and hydrology were taken into account. Garfi and Bruno (2007) opined that landslides are the outcome of geology, hydrology, topography and morphology, climate and weathering. Anthropogenic parameters such as road contributing area, settlement density, and land use and land cover played are considered as risk elements which played an important role in assessing landslide susceptibility and risk (Mandal and Maiti 2013). Das et al. (2011) suggested geo-technical and geo-environmental factors for landslide studies and he considered stochastic process in landslide studies.

The study of landslides has become much more significant in the present day research of mountain geomorphology. To make the mountain slope more acceptable to human population, the preparation of landslide susceptibility map and introduction of site-specific management options are really needed. In the present work, various factors, i.e. elevation, slope aspect, slope angle, slope curvature, geology, soil, lineament density, distance to lineament, drainage density, distance to drainage, topographic wetness index (TWI), stream power index (SPI), rainfall, normalized differential vegetation index (NDVI) and land use and land cover (LULC) were generated using RS and GIS, and finally their integration has been incorporated using different methods and techniques to develop landslide hazard distribution map. Based on the different levels of landslide, susceptibility developmental activities such as construction of roads, expansion of tea garden area, and extension of settlement area are to be addressed in the concerned study area of the Darjeeling Himalaya. As the Darjeeling Himalaya is a potential place for tourism and where the population density is increasing day by day, the identification and preparation of landslide susceptibility map for regional planning and development is crucial and finally where lies the relevance of the present study.

1.2 Regional Settings of Darjeeling Himalaya

The Darjeeling Himalaya located between Nepal and Bhutan Himalaya belongs to Lower Himalaya. In the North, there exist Sikkim Himalaya and in the south the plains of Bengal. The whole Himalayan region is classified into three broad physiographic units, i.e. the great range of snowy peaks, the lower or outer Himalayas and the sub-Himalayan range. In Darjeeling, the mountains belong to the lower Himalayan section which is a long tortuous range and which is running north to south all along its length. Orientation of mountain range has given birth to several river systems which traverse the whole Darjeeling Himalaya and changes continuously general topography of the region through surface erosion, mass movement, sedimentation, etc. All the river system plays a dominant role in the evolution of landform. Major river systems in Darjeeling Himalaya are the Tista, the Great Rangit, the Little Rangit, the Mahanadi, the Balason, and the Mechi. The continuous growth and development of mountain streams increased the steepness of valley side slope, sharpened the interfluves area, and made the slope more susceptible to landslides. The detailed understanding of morphometric parameters, i.e. relief parameters, shape parameters, and areal parameters of a drainage basin, is essential for the assessment of slope instability at basin scale. In geomorphology, the drainage basin is one of the most important geomorphic systems where all the elements of the basin are linked with each other and any changes on one element may introduce changes of other element. Darjeeling Himalaya is divided into four major geological units, i.e. the Darjeeling gneiss, the Buxa, the Daling series, the Damuda series, the Gondwana the tertiary system. Most of the landslides in Darjeeling Himalaya are controlled by some structural characteristics, i.e. a deep-seated rock wedge (W_Fol × J2) within Daling metasediments, plane failure (P_Fol) along foliation surface (Fol) in sheared phyllonitic rocks and toppling (T_J1) caused by steep-dipping SW-dipping joint plane (J1) within quartzo-feldspathic gneiss (https://www.itc.nl/pdf/newsevents/landslides/vanwesternpresentation.pdf).

The most interesting feature of the geological formation is that the original order of superposition having been completely reversed as a result extensive folding and faulting activities. The extensive compressive forces have resulted into large-scale deformation to all the lithological unit and developed cracks, joints, thrust plane, and schist plane throughout the region. The deformed lithology whenever comes into contact with heavy downpour it reduces cohesion which promotes slope failure. The study area is located in tropical climatic region. Rainfall occurs in the month of June, July, and August when the probability landslides is very high. It was observed that most of the destructive landslide events took place during rainy period. The months of November and December are almost rainless when there is no probability of landslide events. The rainfall varies from one slope segment to another segment as a result of variation of elevation, slope aspect, and topographic obstructions. Basically, the southern escarpment slope of Darjeeling Himalaya is attributed with maximum down pour and southern, southeast, and southwest facing slope are also characterized by high landslide susceptibility (Fig. 1.1).

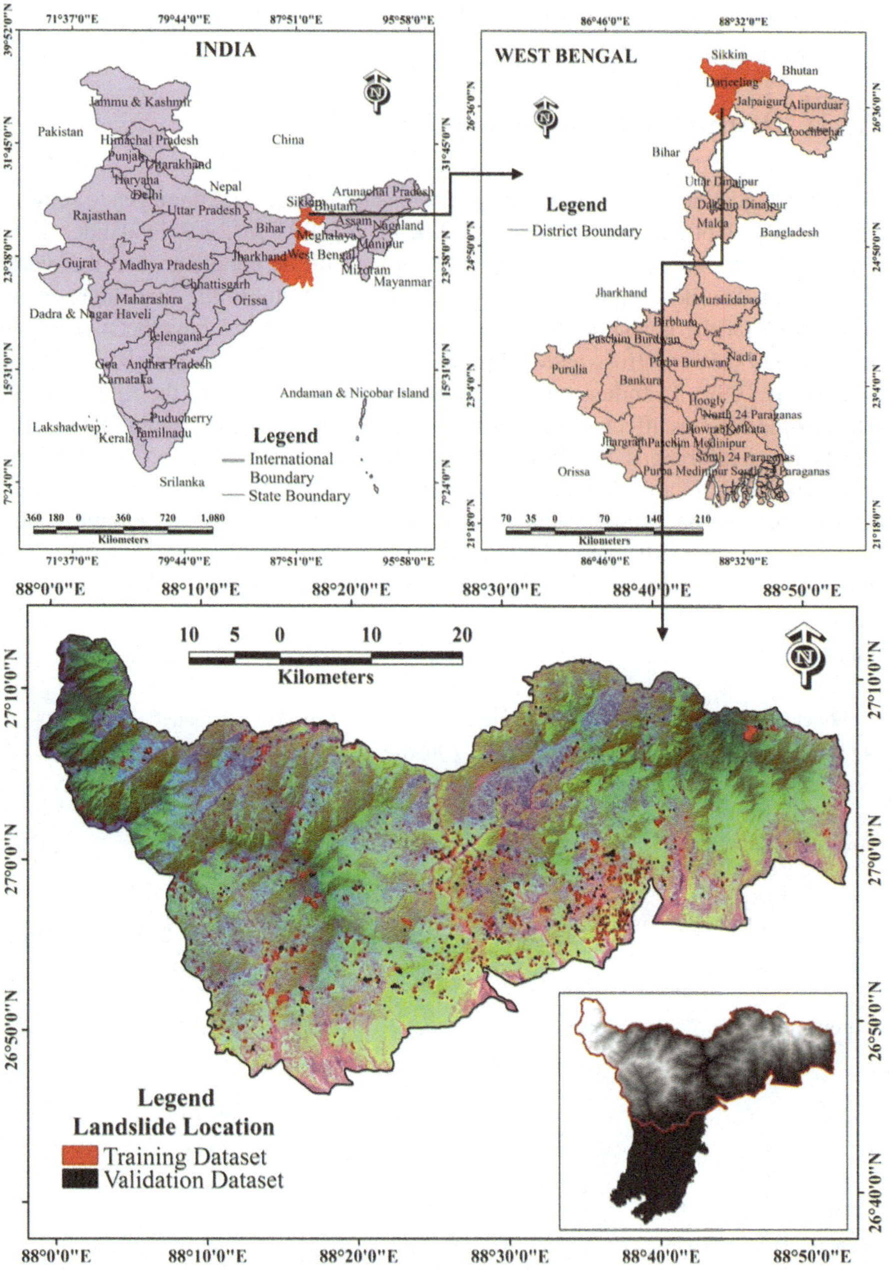

Fig. 1.1 Location map of Darjeeling Himalaya and distribution of landslides

1.3 A Brief Review on Landslide Distribution: World and India

1.3.1 Destructive Landslides of the World

Globally, landslides cause thousands of deaths and injuries and damages to billions of dollars (Aleotti and Chowdhury 1999). From 1903 to 2007 landslides have caused 57,028 deaths and affected more than ten million people around the world (Table 1.1). The developing countries suffer the most due to the landslides event. About 95% of the landslide disasters in developing countries have been recorded and per year 0.5% of gross national product has been lost to the landslides (Chung and Fabbri 1999). In Indian scenario, landslides mostly occur in the rainy season. The most landslides affected areas of India are the Himalayas mountain regions, the Northeastern hill ranges, the Western Ghats, the Nilgiris, the Eastern Ghats, and the Vindhya. All these are mountainous regions in which landslides are the significant natural denudation process and caused by several factors like geology, relief, morphology, etc. and triggered by rainfall, earthquakes, rapid stream erosion, storm waves, etc. (Naranjo et al. 1994; Dai et al. 2002). The occurrences of landslides over

Table 1.1 World scenarios of landslides statistics

Continents		No. of events	Died	Injured	Homeless	Affected	Total affected	Damage US (000'S)
Africa	Total	23	745	56	7936	13,748	21,740	N.A
	Avg./event		32	2	345	598	945	NA
America	Total	145	20,684	4809	186,752	4,485,037	4,676,598	1,226,927
	Avg./event		143	33	1288	30,937	32,252	8462
Asia	Total	255	18,299	3776	3,825,311	1,647,683	5,476,770	1,534,893
	Avg./event		72	15	15,001	6462	21,748	6019
Europe	Total	72	16,758	523	8625	39,376	48,524	2,487,389
	Avg./event		23	7	120	547	674	34,547
Oceania	Total	16	542	52	18,000	2963	21,015	2466
	Avg./event		34	3	1125	185	1313	154
Total		511	57,028	9216	4,046,624	6,188,807	10,244,647	5,251,675

Source: EM-DAT database (1903–2007) (OFDA/CRED 2007) Mountain topography all over the world is attributed with landslides phenomena. There exists variation of topography, climate, soil, natural vegetation, lithology, etc. and as a result of which causative factors, types, and mechanism of landslides also vary from one region to another over the earth surface. Countries of China, India, Japan, Nepal, and Taiwan of Asia; Italy, Germany, and Great Britain of Europe; Canada and the USA of North America; Brazil, Chile, Colombia, and Equador of South America; and Australia and New Zealand of South Pacific are dominated by landslide activities (Table 1.2) where the information and distribution of landslides vary from national scale to catchment and regional scale

Table 1.2 National, regional, and catchment scale landslide studies in the world

Continent	Country affected mostly by landslides	Type	Scale	References
Africa	Nigeria, South Africa	Information and distribution	Catchment and regional scale	Schoeneich and Bouzou (1996)
Asia	China, India, Japan, Jordon, Korea, and Taiwan	Information, distribution, susceptibility, inventory	Catchment and regional scale	Tang and Grunert (1999); Sarkar (1999a, b); Gupta (2000); Kim et al. (2001); Chang and Slaymaker (2002); Mandal and Maiti (2013)
Europe	Italy, Germany, Great Britain, Spain, the UK, and Sweden	Inventory and distribution	Catchment and regional scale	Panizza et al. (1996); Jonasson et al. (1997); Cuesta et al. (1999); Lee and Clark (2000); Schmidt and Beyer (2001, 2003); Hutchinson and Bromhead (2002)
North America	Canada, Puerto Rico, the USA	Inventory	Catchment and regional scale	Brabb et al. (1989); Larsen and Torres-Sanchez (1998); Brardinoni et al. (2003)
South America	Brazil, Chile, Colombia, Ecuador	Inventory and distribution	Catchment and regional scale	Erickson et al. (1989); Martinez et al. (1995); Tibaldi et al. (1995); Amaral and Palmeiro (1997)
South Pacific	Australia and New Zealand	Inventory and distribution	Catchment and regional scale	Page et al. (1994); Glade (1997); Twidale (2000)
Africa Asia	Ethiopia, China, India, Iran, Japan, Jordon, Korea, Nepal, Ukraine	Landslide susceptibility assessment	Catchment and regional scale	Aniya (1985); Bhattacharya (1999); Kamai et al. (2000); De Jaeger (2000); Uromeihy (2000); Ayalew (2000); Anbalagan et al. (2000); Meng et al. (2000); Dhakal et al. (2000); Cherkez et al. (2000); Lee et al. (2001); Dai and Lee (2001, 2002); Lee et al. (2002)
Europe	Austria, Germany, Italy, Portugal, Spain	Landslide susceptibility assessment	Catchment and regional scale	Carrara et al. (1977); Carrara (1989); Tavares and Soares (2002); Fabbri et al. (2002); Fernandez-Steeger et al. (2002)
North America	Canada, USA, Jamaica, Argentina	Landslide susceptibility assessment	Catchment and regional scale	Maharaj (1993); Schmidt et al. (2001); Moreiras (2004); Holm et al. (2004)
South America	Brazil and Colombia	Landslide susceptibility assessment	Catchment and regional scale	Barros et al. (1991); Chung et al. (1995)
South Pacific	Australia and New Zealand	Landslide susceptibility and hazard assessment	Catchment and regional scale	Hayne and Gordon (2001); Glade (2001)

the hilly landscape of the Darjeeling Himalaya are the most serious concern which creates constant problems and causing serious damages to properties and lives as well as hampering the communication line in every year. However, the process has been accelerated by extensive deforestation, unsystematic and unscientific constructional activities, inadequate drainage system, urban expansion, rapid changes of land use practice, etc. In this region, landslides commonly occurred on cut slopes, embankments alongside roads and highways, and also occurred near the high-rise apartments and residential areas (Cheng et al. 2008).

1.3.2 Destructive Landslides of Twenty-First Century in India

Twenty-first century received some destructive landslides all over the world such as Ambooria Landslide in Kerala, India (2001); Chittagong, Bangladesh (2007); Cairo, Egypt (2008); Madeira Island, Portugal (2010); Bududa District, Uganda (May, 2010); Jiang Zhidong Jiangxi, China (May, 2010); Meager Creek, British Columbia, Canada (August, 2010); Iron County, Utah, United States (October, 2011); Salt Lake City, Utah, United States (April, 2013); Kedarnath, Uttarakhand, India (June, 2013); Rockville, Utah, United States (13 Dec 2013); Oso, Washington, United States (22 Mar 2014); Argo District, Badakhshan Province, Afghanistan (2 May 2014); Sunkoshi, Malin, Ambegaon taluka, Pune district, Maharashtra, India (30 Jul 2014); Sindhupalchok District, Nepal (2 Aug 2014); Hiroshima Prefecture, Japan (20 Aug 2014); Badulla District, Sri Lanka (29 Oct 2014); Badakhshan Province, Afghanistan (23 April, 2015); Salvador Bahia, Brazil (28 April, 2015); Lidong Village, Zhejiang, China (13 November, 2015); and Mocoa, Colombia (2 April, 2017) [Source: https://en.wikipedia.org/wiki/list_of_landslides].

In 1950, Darjeeling Himalaya faced destructive landslide events which affected Darjeeling Town, Kalimpong, Takdah, Kurseong, Happy valley, and Tindharia. Another massive destruction took place in 1968 due to landslides which inflicted great damages to human lives and properties at Darjeeling Town, Manipuri, Tista Bazar, Lebong, and Kalimpong. Several tea garden areas were being destroyed completely in these landslide phenomena. In 1980, a massive destruction caused as a result of landslides which killed 215 people and destroyed Hill Cart Road and NH-31. In the last four decades, Darjeeling Himalaya confronted several landslide events year, i.e. 1976, 1991, 1993, 2003, 2005, 2006, 2009, 2011, and 2015 (Figs. 1.2, 1.3, and 1.4). Most of the landslides occurred in Darjeeling Himalaya caused due to continuous showers for few days or short-span high intensity rainfall. Agarwal and Narain (1991); Sarkar (1999a, b); Starkel and Basu (2000); Basu and De (2003); Basu et al. (2009); Chakraborty et al. (2011); The Hindu 2015 revealed that the destruction of 2011 landslides resulted from heavy rainfall in Darjeeling (152 mm), Kalimpong (60 mm), and Kurseong (60 mm) and snatched many lives and properties.

A destructive landslide event took place in Pune Village on 30th July, 2014, which was 110 Km. from Pune city and where 20 people were killed and 150

Fig. 1.2 Destruction of bridges in Kalimpong, 2015 [https://thehimalayantimes.com]

Fig. 1.3 Destruction of human settlement at Darjeeling July, 2015 http://dunyanews.tv/en/World

Fig. 1.4 Destruction of human settlement at Darjeeling July, 2015 [http://kalimpongnewsexpress.blogspot.in]

missing. This very landslide was triggered by heavy rain which flattened the village totally. Twenty dead bodies were being recovered from the heap of mud and debris which moved downward from the nearby hillock. Almost entire tribal village having 50 families were destroyed totally. Only 20 people rescued from the debris who were seriously injured. Finally, there was no sign of human life [http://indianexpress.com]. A series of destructive landslides took place which were aggravated by heavy rain and killed 38 people. This destruction also caused 23 missing and more than 500 displaced in Darjeeling Himalaya, West Bengal. The landslides at 25 places in Darjeeling, Kalimpong, and Kurseong sub-divisions affected badly the National highways (NH-55 and NH-10), including a bridge and complete cut-off of the communication lines. According to the state disaster management department, 22 deaths occurred in Mirik, where 13 people were reported missing. The remaining 17 died in Kalimpong, Lava, Sukhia, and Gorubathan areas. Twenty persons were reported injured and 15 missing at 8 mile and 11 mile areas in Kalimpong [Source: htpp://indianexpress.com] (Figs. 1.5, 1.6, and 1.7). In 2016, a devastating landslide event took place in Tawang in Arunachal Pradesh where 16 persons who were all construction labourers and they were killed (Figs. 1.8 and 1.9). Two persons on the other hand escaped. Tawang is an important tourist destination and Buddhist pilgrimage centre of the eastern Himalaya. Officials in the Tawang district administration said 16 persons were killed when a massive landslide buried the tin-shed house

Fig. 1.5 Destruction of road at Kalimpong July, 2015 [http://kalimpongonlinenews.blogspot.in]

Fig. 1.6 Destruction of road at Kalimpong (Tingling Village) July, 2015 [http://blogs.agu.org]

in which they were staying. While the incident occurred at around 3:30 AM when the labourers were in deep sleep, rescue operations began only at around 5 AM [http://indianexpress.com]. The landslide occurred on July 30, 2014, in a village in Malin (Figs. 1.10 and 1.11). A landslide in the Malin village of Ambegaon Taluka in Pune on Wednesday that killed at least 30 people is an alarming sign for houses that are located in and near hilly areas. In the last 25 years, India has witnessed hundreds such incidents killing thousands of people. But we yet fail to learn any-

Fig. 1.7 Destruction at Kalimpong, July 2015 [http://kalimpongonlinenews.blogspot.in]

Fig. 1.8 Landslide in Tawang Village, Arunachal Pradesh, 2016

thing from these past incidents. Every year in India, especially in the mountain regions like Himalayas, landslides damage and destroy many houses, and cause millions of rupees damage to buildings, roads, railways, pipelines, agricultural land, and crops [http://www.dnaindia.com].

In India, all the landslide activities are triggered by rainfall which leads to elevation of pore-water pressure and seepage. All the mountain terrains in India are attributed with vibration caused due to earthquakes. Such tectonic activity promotes disintegration of lithological unit and development numerous cracks, joints, and

Fig. 1.9 Tawang landslide, April 2016. Photo: Indian Army [http://floodlist.com]

Fig. 1.10 Malin landslide, Maharashtra, [http://www.swadeshnews.com]

Fig. 1.11 Malin landslide, Maharashtra [http://www.swadeshnews.com]

weaknesses plane. Himalayan mountain region is composed of very fragile and weak lithology which is characterized by heavy compression and disintegration. Such fragile lithology whenever comes into contact with heavy showers invites more destructive landslide processes. Beside this, the undercutting of the base of the steep slope and loss of basal support as a result of fast-flowing mountain stream may leads to destructive slope failure. Not only, the natural processes are responsible for destructive landslides in India, but also some significant anthropogenic processes such as deforestation, modification of slopes by construction of human infrastructure for development, mining and quarrying activities, excavation or displacement of rocks, modification of natural drainage system, leakage of pipes such as water and sewer reticulation, and heavy shocks and vibration caused due to plying of heavy loaded vehicles and blasting are also causing great damage over mountain slope and finally introducing slope instability. The *inventory of landslides* includes location, classification, volume, activity, and the date of occurrences of landslides in a geographical region. Landslide inventory map is being prepared with the help of aerial photographs, satellite images, previous landslide records, and Google earth images. *Landslide susceptibility* is the study to assess spatial distribution of landslides which exist or potentially may happen in a region. It also includes velocity and intensity of landslides. It involves the classifications of potentially landslide area.

1. Jul 2, 2016, 29 Dead, Many Missing After Cloudburst In Uttarakhand [www.ndtv.com/topic/landslides-in-uttarakhand]
2. Jul 1, 2016, 10 Killed, Many Missing After Cloudburst, Landslides In Uttarakhand [www.ndtv.com/topic/landslides-in-uttarakhand] (Fig. 1.12)
3. In 2016 (during winter), Rangpo of Kalimpong which is on NH-10 faced a landslide event and four vehicles were buried in the slide but there was only one fatality (Fig. 1.13).

Fig. 1.12 Landslide in Nainital, Uttarakhand, 2016 [http://post.jagran.com]

Fig. 1.13 Landslide at Rangpo in 2016 [http://savethehills.blogspot.in]

4. At least 16 people died in landslide in Phamla village in Tawang district on Friday, 22 April 2016. A further landslide in Tawang district killed at least two people and destroyed property on Saturday 23 April [http://floodlist.com]
5. Jul 11, 2015, Heavy Rain Lashes North India, Parts of Delhi Waterlogged [www.ndtv.com/topic/landslides-in-uttarakhand]

Fig. 1.14 Landslide in Darjeeling Town, 2011 [http://savethehills.blogspot.in]. Intense thundershowers from approx. 6 pm on 24 Aug 2011 dumped approx. 130 mm of rainfall over a 5 h period in Darjeeling and resulted in this landslide at the Mall, Chowrasta, which is in the heart of the town in the early hours of the morning today

6. July, 2015, 30 people were killed after a series of landslides that triggered by overnight torrential rain and hit different parts of the Darjeeling Hills.
7. Malin landslide, Maharashtra: The landslide occurred on July 30, 2014, in a village in Malin. The landslide occurred due to heavy rainfall and around 151 people died and 100 people went missing after the disaster [http://indiatoday.intoday.in]
8. June, 2013, a destructive landslide took place in Kedarnath which took at least more than 5700 casualties.
9. September 24, 2012, Northern Sikkim a destructive landslide took place which invited a death toll of 27 people including the members of the Indo-Tibetan Border Police.
10. August 25, 2011, faced a destructive landslide in the heart of the Darjeeling Town (Fig. 1.14)
11. July 27, 2007, at Dasalgaon of Maharashtra more than 50 casualties occurred due to landslide events.
12. July 26, 2005, at Raigad in Maharashtra 54 people were killed due to destructive landslides.
13. July 26, 2005, another landslide at Sakinaka of Mumbai took away more than 74 people.
14. In November 2001, 40 people were killed at Amboori in Kerala due to landslide.

15. Mumbai landslide, Maharashtra: The landslide was caused in July 2000. The landslide took place in the suburbs of Mumbai due to heavy rains which was followed by land erosion. As per reports, around 67 people died and the local trains were also stricken [http://indiatoday.intoday.in]
16. August, 1998, Malpa willage was completely wiped out which caused approximately 400 casualties.
17. August 14, 1998, 69 people were killed at Okhimath due to landslides.
18. June, 1994, in Kanka Coast 20 people were killed and breaching of ghat also caused damages of road around 1 km.
19. August, 1993 in Nagaland 200 houses were destroyed and 500 people died which caused heavy loss to human lives and properties.
20. August, 1993, Kalimpong faced a destructive landslide phenomenon and took away about 40 people which led to heavy loss of property.
21. July, 1993, at Itanagar a destructive landslide took place which inflicted killing of 25 people who were buried alive.
22. July, 1991, in Assam 300 people were killed, road and buildings were damaged.
23. Darjeeling landslide, West Bengal: The landslide happened around October 4, 1968. The landslide was triggered by floods and the 60 km long highway was cut off in 91 parts. As per reports, thousands of people died in the landslide [http://indiatoday.intoday.in]
24. Guwahati landslide, Assam: The landslide took place on September 18, 1948, due to heavy rains. Over 500 people died in the landslide and according to the reports, the landslide buried an entire village [http://indiatoday.intoday.in].

1.4 Landslides: Types, Causes, and Mechanism

Landslides are the down slope movement of rock-soil debris under the influence of gravity. The formation of debris is the result of disintegration and decomposition of rocks by physical and chemical weathering processes. The movement of disintegrated and decomposed slope material occurs when driving force of the slope soil exceeds the resisting forces. Various causative factors such as slope, relief, curvature of the slope, physical and chemical properties of soil, lithology, natural vegetation, rainfall intensity, earthquake, existence of weaknesses planes, and tectonic activities play an important role in making the mountain slope more vulnerable to landslides. The rate of movement of the slope materials varies from slow to rapid and on the basis of which landslide can also be classified into *fall*, *slide*, *slump, creep,* etc. The displacement of the debris occurred in different direction, i.e. *vertical*, *diagonal,* and *lateral,* and the mass movement can be treated as vertical mass movement, diagonal mass movement, and lateral mass movement. Movement of mass in different direction over the mountain slope helps to increase the areal extent of the landslides. The movement of slope materials are governed by several factors, i.e. *predisposing factors* which includes the physical and chemical properties of soil and rocks; *preparatory factors* denotes the reduction of soil strength by weathering

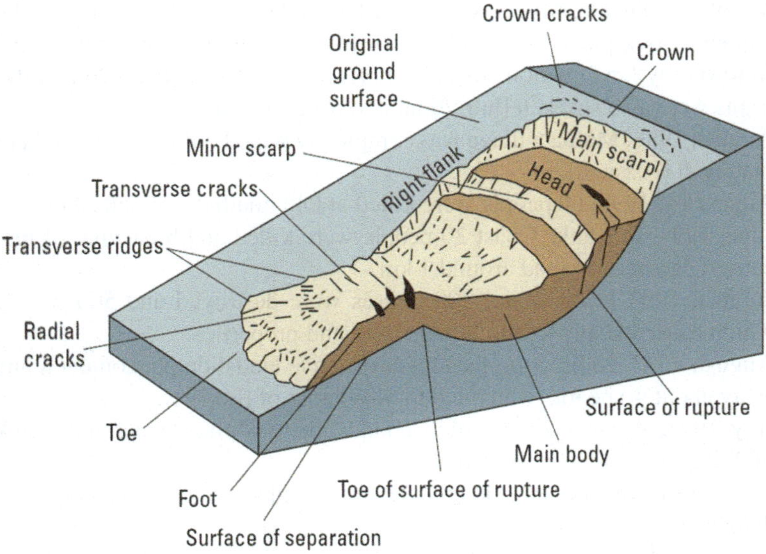

Fig. 1.15 A simple illustration of rotational landslide (Highland and Bobrowsky 2008)

(Chandler 1972), climate change (Dehn et al. 2000), and tectonic uplift (Shroder and Bishop 1998); *triggering factors* which promotes movement of slope materials from a marginally stable to an actively unstable state and governed by intense rainstorms, prolonged wetted condition, seismic tremor and undercutting of the slope and *sustaining factors* which indicate the character of actively unstable slope and influenced by rainfall.

Rotational landslides are characterized by a number of diagnostic features such as ground surface, crown cracks, crown, surface of rupture, major scarp, minor scarp, head, main body of the slide, toe of surface of rupture, surface of separation, transverse cracks, transverse ridges, radial cracks, toe and foot (Fig. 1.15).

1.4.1 Types of Landslides

Landslides are closely associated with the fluvial system. Continuous and regular growth of drainage network over the steep mountain slope steepens the slope, and concentration of drainage over the space promotes soil saturation. The development of first- and second-order streams in the upstream area and their active erosion provide large amount of materials to the downstream areas which are being deposited valley side slope. Such deposition of sediments move downwards whenever it increases its weight die to wetting and adding more materials from upslope and decreases shearing strength. Korup (2003) introduced coupling model where he ascribed the geometric relationship between landslide and fluvial system. He opined that large landslides having an areal extent of 100 km^2 eliminate low-order drainage divides and reorient the drainage system. Most of the run-out direction of the landslide corresponds to the drainage system which is followed by general slope of the land. Valley side landslides have direct impact on morphological system of the channel. Such event may change hydro-geomorphological character of the channel. Landslides are deposited in the alluvial fan zone, reservoirs, lakes, and floodplain areas which can contribute in the formation of terraces, fans, floodplain alluvium, valley fill, etc. A glacial process also plays a significant role in the down slope movement of slope materials. Processes of *frost shattering*, *frost heaving*, *frost cracking*, and *solifluction* and *congeliflutio*n have strong influence on loosening of slope materials and their movement in the form of landslides. Glacial debris with ice particles move faster than any other kind of materials existing over the mountain slope where ice particles act as a lubricating agent. Frost shattering and frost cracking help in the formation of rock block and introduce rock fall. It is clear that fluvial and glacial processes which control geomorphic system variables and contribute a lot to promote landslides in the mountain environment. Vernes (1978) identified falls, topples, slides, lateral spreads, and flows (Table 1.3) which were then classified into several types based on slided materials. Considering the materials (rock, debris, and earth) existing in the slide, Dikau et al. (1996) classified landslides into eight types, i.e. fall, topple, rotational slide, translational slide, planar, lateral spreading, flow and complex (Table 1.4).

Site-specific analysis of rock slope failure, rock fall and topple were summarized by Evans and DeGraff (2002). They used simple empirical methods for measuring fall height and volume of the rocks to assess the run-out distance. Monma et al. (2000) opined that the detail study on rock failure requires more sophisticated models. Various conventional methods, i.e. stereographic and kinematic analysis (determination of critical slope, discontinuity geometry, and shear strength characteristics), limit equilibrium analysis (determination of the nature of stability assessing slope geometric and materials characteristics and rocks shear strength characteristics), and physical modelling (models based on slope geometry, rocks

Table 1.3 Five major types of landslides

	Slides materials		
		Engineering soils	
Type of movement of slides	Bedrock	Predominantly coarse	Predominantly fine
Falls	Rock fall	Debris fall	Earth fall
Topples	Rock topple	Debris topple	Earth topple
Slides (rotational and translational)	Rock slide	Debris slide	Earth slide
Lateral spreads	Rock spread	Debris spread	Earth spread
Flows	Rock creep	Debris creep	Earth creep

Source: Vernes (1978)

Table 1.4 Classification of landslide after Dikau et al. (1996)

	Materials		
Process	Rock	Debris	Earth
Fall	Rock fall	Debris fall	Earth fall
Topple	Rock topple	Debris topple	Earth topple
Rotational slide	Slump	Single, multiple, and successive	Single, multiple, and successive
Translational slide	Block slide	Block slide	Slab slide
Planar	Rock slide	Debris slide	Mud slide
Lateral spreading	Rock spreading	Debris spread	Earth spreading
Flow	Rock flow (Sackung)	Debris-flow	Earth flow
Complex	Rock avalanche, Bergsturz	Flow slide	Slump-earth flow

sizes, shapes, etc) were applied to analyse rock slope failure. All these conventional methods were studied by Stead et al. (2001), Moser (2002), Jones et al. (Colorado Rockfall Simulation Model CRSM in 2000), Spang and Sonser (1995), Evans and Hungr (1993), and Guzzetti et al. (2002). Numerical models in rock slope failure studies include continuum modelling, discontinuum modelling (Yamagami et al. 2001), and hybrid/coupled modelling (Moser 2002).

Amongst various mass movements, debris-flow is one of the most complex processes (Hungr 2005). This kind of slope failure is quite common in Darjeeling Himalaya. The studies on debris-flow made by Chen (1997), Wieczorek and Naeser (2000), and Rickenmann and Chen (2003). The assessment of run-out distance is required to understand the character of debris-flow of a region (Wieczorek and Naeser 2000). Debris-flow analysis also considered the volume of rock-waste, relief, and travel distance covered which can be analysed using the empirical models. Debris-flow study is being accomplished with the help of detailed information on rheologic, hydrologic, and hydraulic properties (Coussot et al. 1998). Both statistical and numerical approaches were being applied to study debris-flow worldwide (Mark and Ellen 1995; Lorente et al. 2002).

1.4.2 Mechanism and Causes of Landslides

Landslides are being regulated by a particular type of natural processes which are also the resulted outcome of slope materials characters, slope steepness, and degree of disruption of the mass. Most of the researchers classified landslides based on the materials present over the slope, processes, and rate of movement of the materials. Landslide influencing factors can be divided into two, i.e. permanent and variable (Sharpe 1938). The permanent factors are slope steepness and rock type. After thorough examination of a specific landslide phenomenon of an area, it becomes possible to identify how permanent factors are responsible for landslide processes. These factors are not being changed frequently but could be changed only after a long period of time. Variable actors include vibration of ground due to seismicity, changes of the ground water level, increase in the soil moisture conditions as a result of high intensity precipitation, etc. Both, permanent and variable factor played an important role in landslide processes. To promote landslide in a geographical region, *past landslide and its distribution*, *bed rock*, *slope steepness* or *inclination*, *hydrological factors,* and *human intervention* were considered as significant contributing factors. The location of past landslide area can be reactivated and may again change the hydro-geomorphologic character and may initiate landslip again. The site-specific analysis of past landslide and associated geomorphic, hydrologic, and geo-technical parameters indicate a particular location as susceptible to landslides or not. Human intervention over the slope such as construction of road and settlement and elimination of the forest cover may change slope steepness and ground water situation which may induce slope instability (Swanson and Dyrness 1975). The occurrences of landslides phenomena cause great damage to human lives and properties worldwide. Rapid rate of urbanization and concentration of human settlements and associated developmental activities, i.e. construction of roads, dams, expansion of the settlement area, clearing of the slope vegetation over the steep mountain slope have aggravated the problems of landslide hazards and disasters. The hill slope stability can be expressed in terms of the relationship between the forces tending to drive the hill slope materials and the forces tending to resist driving stress (Table 1.3). When driving force exceeds the resisting force, the slope materials will start to move down the slope under the influence of gravity. This relationship can be expressed in terms of safety factor (FS).

$$\mathrm{FS} = \frac{\mathrm{ss}}{\tilde{i}} \quad \text{(Eq. 1.1)}$$

where ss is the total shear strength along a specific shear plane and $\tilde{i}$ is the total amount of shear stress developed along this plane.

Landslides are the expression of the relationship between shear stress and strength. Both shear stress and shear strength are being influenced by cohesion, angle of internal friction, shear surface, gravitational force, normal stress, and slope angle of the slope surface (Fig. 1.16). All these parameters are the result of weathering, wetting, and swelling of the soil, ground water level, saturation of the slope materials, removal of the underlying support, increase in the weight of slope materials, pore-water pressure, etc. (Table 1.5).

The causes of landslides are usually related to slope materials movement processes. There are several causes and factors which promote landslide phenomena (Table 1.6). Sometimes, the terms causes and trigger create confusion. Causes may be considered as factors that make slope becoming unstable or vulnerable. On the other hand, trigger is the single event that finally initiates the movement.

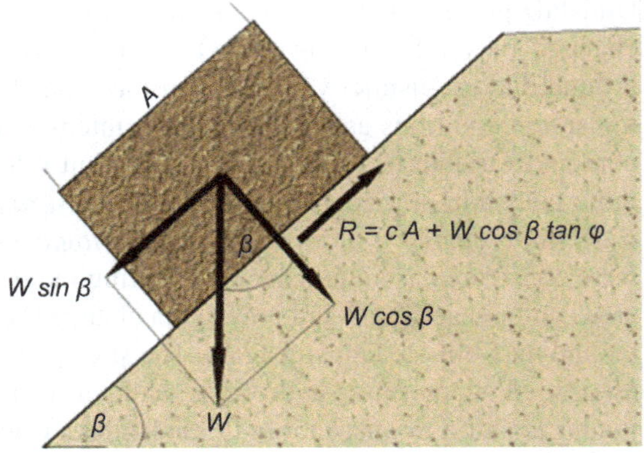

Fig. 1.16 Schematic illustration of the forces acting on the slope which determines safety factor (F) [where A-shear surface; W-gravitational force; $W \sin \beta / A$-shear stress; $s = c + \sigma \tan \varphi$ [shear strength]; σ = normal stress [$W \cos \beta / A$]; c = cohesion]. (**Source**: Caribbean Handbook on Risk Information Management)

Table 1.5 Factors of increase of shear stress and decrease of shear strength

Factors which increase the shear stress	Factors which decrease the shear strength
Removal of lateral support by rivers, glaciers, sea wave, fault, previous rock fall, or slide	Disintegration and decomposition of rocks due to weathering, hydration process, etc.
Removal of underlying support by rivers, sea wave, etc.	Saturation and weakening of weathered materials, alter in pore-water pressure
Slope loading due to the weight of vegetation, water, and debris	Changes of structure due to the remoulding of sands, creation of fissures in shales and clays, etc.
Lateral pressure due to the water present in the cracks, freezing in cracks, swelling and pressure release, etc.	Organic impact in respect of burrowing of animals and dead organ of tree roots
Transient stress due to the earthquake	

Source: Vernes (1978)

1.5 Objectives of the Present Study

Darjeeling Himalaya exhibits a wide range of geomorphic and relief diversity. The study of various factors and location of landslides and preparation of landslide susceptibility map with reasonable accuracy with the help of RS and GIS and statistical methods have become much more significant in the present day. The present study is dealt with the application of various statistical methods in landslide susceptibility assessment and prediction of Darjeeling Himalaya, India. Here, a comparison of various statistical approaches was made to find out most suitable method in landslide studies. Major objectives of the present work are:

- Identification of different causative factors responsible for landslide phenomena in Darjeeling Himalaya and the preparation of data layers corresponding to the factors.

Table 1.6 Landslide causes and triggering mechanisms

	1. *Ground conditions*
1	Plastic weak material, sensitive material, collapsible material, weathered material, sheared material, and jointed or fissured material
2	Adversely oriented mass discontinuities (including bedding, schistosity, cleavage) and structural discontinuities (including faults, unconformities, flexural shears, sedimentary contacts)
3	Contrast in permeability and its effects on ground water contrast in stiffness (stiff, dense material over plastic material)
	2. *Geomorphological causes*
1	Tectonic and volcanic uplift
2	Glacial rebound
3	Fluvial, glacial, and wave erosion of the slope toe
4	Erosion of the lateral margins and subterranean erosion (solution, piping)
5	Deposition loading of the slope or its crest
6	Vegetation removal (by erosion, forest fire, drought)
	3. *Physical causes-triggers landslides*
1	Intense, short period rainfall and prolonged high precipitation
2	Rapid melt of deep snow
3	Rapid drawdown following floods, high tides, or breaching of natural dams
4	Earthquake and volcanic eruption
5	Breaching of crater lakes and shrink-and-swell weathering of expansive soils
6	Thawing of permafrost, freeze-and-thaw weathering
7	Freeze-and-thaw weathering and shrink-and-swell weathering
	4. *Man-made causes for landsliding*
1	Excavation of the slope or its toe and loading of the slope or its crest
2	Drawdown (of reservoirs) and irrigation
3	Defective maintenance of drainage systems and water leakage from services (water supplies, sewers, storm water drains)
4	Vegetation removal (deforestation) and mining and quarrying (open pits or underground galleries) and creation of dumps of very loose waste
5	Artificial vibration (including traffic, pile driving, heavy machinery)

Source: Highland and Bobrowsky (2008)

- Application of geo-spatial tools and statistical models to integrate various data layers and to prepare landslide susceptibility map of Darjeeling Himalaya.
- Comparison between frequency ratio (FR) model, modified information value (MIV) model (MIV), logistic regression (LR) model, analytical hierarchy process (AHP) model, artificial neural network (ANN) model, weighted overlay analysis (WOA) model, certainty factor (CF) model and fuzzy logic (FL) approach based on validation and accuracy results to identify the suitable technique for landslide susceptibility assessment studies in Darjeeling Himalaya.

1.6 Data Sources for Landslide Susceptibility Assessment Studies

To accomplish the present work, various secondary data were applied carefully. To validate the result, field investigation with GPS conducted to identify landslide locations in Darjeeling Himalaya. Historical landslide records and others relevant information were obtained from research articles (Starkel 1972; Basu and Sarkar 1985; Basu and Ghatwar 1988; Basu and Maiti 2001; Maiti 2007; Ghosh et al. 2009), newspapers (Ananda Bazar Patrika-10/09/06, 14/06/2007, 16/07/2007 and 18/07/2007, Telegraph-04/08/2007, Statesman-August, 2006), and Geological Survey of India Technical Report. Various data such as Google earth image (2015–2016), map of geology (GSI, Kolkata) and soil (NBSS & LUP), SRTM DEM (30 m*30 m), and Landsat TM Image (2009) were used to prepare landslide inventory map, geology and soil map, slope angle, slope aspect, slope curvature, SPI, TWI, lineament density, distance to lineament, distance to drainage, drainage density and NDVI map of Darjeeling Himalaya (Table 1.7). All the data layers were made on ARC GIS environment.

Table 1.7 Data type, data source, and its application

Purpose	Data used	Data source	Data type
Administrative boundary of India and West Bengal	Map of census of India, 2011	http://www.censusindia.gov.in/2011census/maps/atlas/00part1.pdf	ARC/INFO polygon coverage
Preparation of landslide inventory map	Google earth imagery, toposheet, and GPS field survey data (2016–2017)	Google earth map, Survey of India (SOI), Kolkata. Toposheet no. 78A/4, 78A/8, 78A/12, 78A/16, 78B/1, 78B/2, 78B/5, 78B/6, 78B/9, and 78B/13 GPS data	ARC/INFO polygon
Thematic data layer of geology and soil types	Geological and soil map	Geological Survey of India, Kolkata NBSS and LUP Regional Centre, Kolkata	ARC/INFO polygon coverage
Preparation of altitude aspect, slope, curvature, SPI, and TWI/CTI thematic data layer	SRTM DEM (30 m spatial resolution)	earthexplorer.usgs.gov	ARC/INFO grid
Preparation of LULC, lineament density, and distance from lineament data layer	Land use map, drinking water prospects map	www.banglarbhumi.gov.in, www.wbphed.gov.in	ARC/INFO polygon
Thematic data layers of drainage density and distance to drainage	Disaster mitigation map of Darjeeling district	http://darjeeling.gov.in/gismaps.html	ARC/INFO polygon
Preparation of rainfall distribution data layer	Rainfall data (1950–2010)	http://www.worldclim.org	ARC/INFO grid
Data layer of NDVI	Landsat TM Image	earthexplorer.usgs.gov	ARC/INFO grid

1.7 Landslide Susceptibility Assessment: Techniques and Methods

Bureau of Indian Standards (BIS-1998) suggested a principle for landslide hazard zonation mapping on 1:25,000 or 50,000 scales. Bhandari (1987) opined landslide hazard zonation map on the basis of hazard rating values in connection to causative factors of landslide. Various attempts have been taken into account all over the world to assess landslides and its destructive character. Carrara et al. (1995); Hutchinson (1995); Soeters and Van Westen (1996); Van Westen et al. (1997); Aleotti and Chowdhury (1999); Guzzetti et al. (1999); and Committee on the Review of the National Landslide Hazards Mitigation Strategy (2004) suggested the most common approaches of landslide susceptibility mapping procedure and which can be grouped into five major categories, namely: (1) direct geomorphological mapping; (2) analysis of landslide inventories; (3) heuristic or index-based methods; (4) statistical methods, including neural networks, fuzzy logic, and expert systems; and (5) process-based conceptual models.

Akgun et al. (2007); Lee and Sambath (2006); Mondal and Mandal (2017) used RS- and GIS-based frequency ratio method for landslide susceptibility mapping. AHP method was used to assess the spatial distribution landside susceptibility by Ayalew et al. (2005); Gorsevski et al. (2006), Wu and Chen (2009); Akgun and Turk (2010); Mondal and Maiti (2012); Kayastha et al. (2013); and Mansouri Daneshvar (2014). Pradhan (2010); Pradhan and Lee (2009); Lee et al. (2007) successfully applied logistic regression method to evaluate landslide susceptibility. Landslide distribution analysis method by Cheng et al. (2004), qualitative/heuristic method by Kanungo et al. (2006), information value (Info Val) method by Saha et al. (2005), landslide normal risk factor method by Doratti et al. (2002), and discriminant analysis by Ardizzone et al. (2002) have been applied to assess spatial distribution of landslides.

In the first decade of twenty-first century, several new susceptibility methods, i.e. neuro-fuzzy method (Pradhan 2010), support vector machine (Xu et al. 2012), decision tree method (Wan 2012), index of entropy (Devkota et al. 2013), Bayesian network (Song et al. 2012), and fractal theory (Li et al. 2011) were incorporated in landslide susceptibility assessment and prediction. In the present study, study 15 landslide conditioning factors, i.e. elevation, slope aspect, slope angle, slope curvature, geology, soil, lineament density, distance to lineament, drainage density, distance to drainage, stream power index (SPI), topographic wetted index (TWI), rainfall, normalized differential vegetation index (NDVI) and land use and land cover (LULC) were incorporated for the preparation of landslide susceptibility zonation maps. Landslide susceptibility assessment techniques can be divided into two main categories, i.e. qualitative method (direct) and quantitative method (indirect). Qualitative method is accomplished with the help of landslide distributions and geomorphic analysis. On the other hand, qualitative approaches include statistical analysis, distribution-free analysis, deterministic analysis, and landslide frequency analysis (Table 1.8).

Table 1.8 Landslide hazard/susceptibility zonation techniques

Methods				References
Qualitative (direct)	Landslide distribution analysis			van Westen and Getahun (2003), Cheng et al. (2004)
	Qualitative/heuristic(geomorphic analysis)			Moeyersons et al. (2004), Kanungo et al. (2006)
Quantitative (indirect)	Statistical analysis for landslide hazard zonation	Bi-variate methods	Probabilistic likelihood (frequency ratio) analysis	Lee and Pradhan (2007), Mandal and Mandal (2016), Mondal and Mandal (2017)
			Information value (Info Val) method	Zezere (2002), Saha et al. (2005)
			Weights of evidence (WofE) method	Sabto (1991)
			Landslide normal risk factor	Doratti et al. (2002), Saha et al. (2005)
		Multi-variate methods	Multiple logistic regression	Lee and Pradhan (2007), Lee et al. (2007), Pradhan and Lee (2009), Pradhan (2010)
			Discriminant analysis	Ardizzone et al. (2002), Baeza and Corominas (2001)
	Distribution-free methods	Artificial neural network (ANN) method		Lee et al. (2007), Choi et al. (2011)
		Neuro-fuzzy method		Pradhan (2010), Tazik et al. (2014)
	Deterministic analysis for landslide hazard zonation			Zorn and Komac (2004), Claessens et al. (2006)
	Landslide frequency analysis			Yu et al. (2006), Giannecchini et al. (2007)

1.8 Statistical Models in Landslide Susceptibility Assessment Studies

The assessment of landslide is required for planning and management as well as development in the Himalayan mountain. Since very beginning, researchers and academicians started to study landslide events in the concerned landslide-prone areas of Darjeeling Himalaya. Earlier, they were relied upon the geomorphological mapping of a landslide location by measuring the length, width, and area of a landslide. They also opined that several factors play an important role in landsliding and considering all the landslide factors and assigning arbitrary weights to each class of individual factor landslide hazard rating value was derived for a particular area. Then based on the total hazard rating value a landslide susceptibility map was made.

The application of RS and GIS and statistical and mathematical modelling has introduced a new dimension in landslide studies. Before the introduction of statistical, mathematical, and RS and GIS, physical process-based model was applied successfully in the study of landslides with poor data base. But at present various bi-variate, multi-variate statistical models are being used with strong data base to assess landslide susceptibility with reasonable accuracy. Landslide can be assessed with the help of several methods such as heuristic, semi-quantitative, quantitative, probabilistic, deterministic, and multi-criteria decision support system approaches. But not all the approaches are affective for the assessment of landslide hazards.

The improvement of geo-informatics along with statistical approaches has opened a new door in landslide susceptibility assessment and prediction. The statistical methods in landslide studies can be classified into two, i.e. bi-variate statistical approaches and multi-variate statistical approaches. Bi-variate statistical approaches include weights of evidence model, weighted overlay method, frequency ratio model, information value model, and fuzzy logic approach. Multi-variate statistical approaches are dealt with basically logistic regression analysis, discriminant analysis, artificial neural network model, etc. Both bi-variate and multi-variate approaches consider data-driven as well as knowledge-driven data and information which helps to perform various model on GIS environment.

Soeters and van Westen (1996) presented the spatial pattern of landslide assessment methods in different scales of resolution such as large scale (<1:10,000), medium scale (1:50,000–1:100,000), regional scale (1:125,000–1:500,000), and national scale (>1:750,000). Qualitative methods in landslide studies follow inventory and heuristic approaches. Inventory approaches provide lot of information with regard to landslides which is helpful for future improvement in spatial analysis of susceptibility assessment (Guzzetti et al. 1999). Heuristic approach is based on prior knowledge, local experience, and expert judgement. This approach applied spatial information which includes topography, hydrology, geology, and geomorphology. Qualitative methods can be used at all scales. Statistical methods are used large scales and medium scales landslide studies (Table 1.9).

Landslides are the major environmental hazards which can be managed with the development of some concrete structures, increasing awareness amongst the people, identifying landslide-prone areas with the help of suitable techniques and introducing

Table 1.9 Scales for different spatial landslide assessment

	Qualitative methods		Quantitative methods		
Scale	Inventory	Heuristic	Statistical	Probabilistic prediction analysis	Process based and numerical analysis
<1:10,000	Yes	Yes	Yes	Yes	Yes
1:15,000–1:100,000	Yes	Yes	Yes	Yes	Probable
1:125,000–1:1,500,000	Yes	Yes	Probable	Probable	No
>1:750,000	yes	yes	No	No	No

Source: Soeters and van Westen (1996)

warning systems (Dai et al. 2002). Varnes (1978, 1984) postulated that landslide susceptibility assessment aims to differentiate a land surface into homogeneous areas according to character of mass movement and propensity of failure. Jibson et al. (2000); Jibson (2000); Rautelal and Lakheraza (2000); Donati and Turrini (2002); and Zhou et al. (2002) suggested various techniques and methods for landslide susceptibility and risk assessment. The present study is dealt with frequency ratio (FR) model, modified information value (MIV) model, logistic regression (LR) model, artificial neural network (ANN) model, weighted overlay analysis (WOA) model, certainty factor (CF) model, analytical hierarchy process (AHP) model and fuzzy logic (FL) approach have been applied to assess landslide susceptibility of Darjeeling Himalaya considering 15 landslide conditioning factors. All these techniques are quantitative as well as knowledge driven. Because, all these approaches are dealt with the assignment of the rating value to each class of all the factors and estimation of class weight as well as factor weight.

1.8.1 Modified Information Value (MIV) Model and Frequency Ratio (FR) Model

Sarkar et al. (2006) applied *modified information value model* (MIVM) successfully to generate landslide susceptibility map which were being followed in the study for landslide assessment and prediction spatially. *Frequency ratio model* is one of the most accepted statistical methods in landslide assessment and prediction. The value of FR is dealt with the relationship between the occurrences of past landslide and the landslide conditioning factors in Darjeeling Himalaya. A detailed investigation with regard to past landslide was made in consultation with Survey of India (SOI) Toposheets, recent Google earth image (2015–2016), satellite images, and research papers to prepare a landslide inventory of Darjeeling Himalaya. The prepared landslide inventory map was integrated with landslide condition factors on GIS platform to assess landslide potentiality of each class of all the data layers. The objective of the study is associated with the preparation of landslide susceptibility map using both frequency ratio model and modified information value model and the estimation of accuracy level using success rate curve. To prepare landslide susceptibility map using frequency ratio model, the study made by Lee and Sambath (2006) and Lee and Pradhan (2006) was consulted in the present research.

1.8.2 Logistic Regression (LR) Model

Logistic regression analysis is one of the most significant statistical methods to find out the relationship between landslide inventory and the geographical data base responsible for landslips (Mancini et al. 2010). This approaches were successfully applied by Dai and Lee (2002); Guzzetti et al. (2006); Lee and Sambath (2006); and Lee and Pradhan (2006). The determination of all the landslide influencing factors and preparation of geographical data set on GIS environment is a vital task to perform linear regression model and prepare landslide susceptibility map. All the data sets were converted to GIS environment (raster value domain) and regression co-efficient value was derived using SPSS software for each category. A landslide inventory of Darjeeling Himalaya was made after intensive field investigation with GPS, satellite images, Google earth images, and historical landslide data which were considered as dependent variable to perform linear regression model. To prepare the landslide susceptibility map of Darjeeling Himalaya, 15 landslide conditioning factors, i.e. elevation, slope aspect, slope angle, slope curvature, geology, soil, lineament density, distance to lineament, drainage density, distance to drainage, stream power index (SPI), topographic wetted index (TWI), rainfall, normalized differential vegetation index (NDVI) and land use and land cover (LULC) were considered. All these factors/data set were taken into account as independent variable in the study. An iterative algorithm was used to find out the non-linear relationship between independent variables and the estimated probability after Dai and Lee (2002). Maximum likelihood method was applied to estimate regression co-efficient. A comparison was made between the result of linear regression model and external data set by random process to validate the methodology after Mancini et al. (2010).

1.8.3 Artificial Neural Network (ANN) Model

Landslide activities in the mountain environment are linked with number of geomorphic and geo-hydrologic variables. Physical-based model cannot properly link between various causative factors of landslide and landslide occurrences phenomena and estimate the spatial distribution of landslide susceptibility properly. To solve the problems of interaction between landslide triggering factors, researchers started to find out a suitable model, and finally they established so many statistical models in landslide studies. *Artificial neural network model* is amongst of them. At present, this model is successfully used to assess and predict landslide susceptibility by various researchers. Artificial neural network model was successfully used by Xie et al. (2004); Pistocchi et al. (2002); Lee et al. (2003); Lee et al. (2004). A landslide distribution map was developed in consultation with toposheet, Google images (2015–16), and field studies and used to estimate the frequency of landslide in the concerned study area. In the present work, ANN model was applied for the assessment of landslide susceptibility of Darjeeling Himalaya, West Bengal. Back-propagation training algorithm was applied on GIS platform to perform the data creation process as well as to develop a network that can provide output value from the input one. Finally, the output values helped to produce the landslide susceptibility map of Darjeeling Himalaya on GIS environment.

1.8.4 Weighted Overlay Analysis (WOA) Model and Fuzzy Logic (FL) Approach

In the *weighted overlay approach*, weighted values are assigned to all the data layers based on relative importance of the factors and their classes which requires an adequate knowledge with regard to landslide phenomena about the concerned region, and finally their integration is being made on GIS environment to prepare landslide susceptibility map. In the present research, a soft computing technique, known as *fuzzy logic*, has been applied to map landslide susceptibility. This approach is dealt with the estimation of fuzzy membership values between 0 and 1. The fuzzy membership values were integrated applying fuzzy gamma operator after Champati ray et al. (2007) which helped to generate landslide susceptibility map of Darjeeling Himalaya. The approach can be regarded as quantitative or semi-quantitative approach. In semi-quantitative approach, the weightings are assigned based on the field knowledge of the experts, whereas in quantitative approaches the weightings are determined according to correlations between landslide locations and landslide conditioning factors. Such an expert-based approach, referred as Trapezoidal Fuzzy Number Weighting (TFNW).

1.8.5 Analytical Hierarchy Process (AHP) Model and Certainty Factor (CF) Model

Analytical hierarchy process (AHP) is a quantitative approach which considered the logical assignment of values against each class and factors to derive class weight and factor weight. This method first proposed by Saaty (1980, 1990, 1994, 2000). In this approach, each class as well as each factor was rated assigning the values 1–9 and a matrix was developed to estimated class weight and factor weight. Finally, a linear algorithm was applied on GIS platform to derive landslide susceptibility co-efficient values and to prepare landslide susceptibility map of Darjeeling Himalaya. In the *certainty factor approach,* certainty factor is derived for each class of all the data layers considering landslide inventory. This approach was first introduced by Shorthliffe and Buchanan (1975) and then by Heckerman (1986). The value of the certainty factor was derived after Binaghi et al. (1998). Then, a linear combination model is being performed on GIS environment to generate landslide susceptibility map of Darjeeling Himalaya. The validation of each approach was made ROC curve and based on the values of area under curve accuracy level was also estimated.

References

Agarwal, A., & Narain, S. (1991). *Floods, flood plain and environmental myths (State of India's Environment: A Citizens' Report, 3)*. New Delhi: Centre for Science and Environment.

Akgun, A., & Turk, N. (2010). Landslide susceptibility mapping for Ayvalik (Western Turkey) and its vicinity by multicriteria decision analysis. *Environmental Earth Sciences, 61*, 595–611.

Akgun, A., Dag, S., & Bulut, F. (2007). Landslide susceptibility mapping for a landslide prone area (Findikli NE of Turkey) by likelihood-frequency ratio and weighted linear combination models. *Environmental Geology, 54*, 1127–1143.

Aleotti, P., & Chowdhury, R. (1999). Landslide hazard assessment: Summary review and new perspectives. *Bulletin of Engineering Geology and the Environment, 58*, 21–44.

Amaral, C., & Palmeiro, F. (1997). Local landslide inventory of Rio de Janeiro: State of the art and access. In ABMS ABGE & ISSMGE (Ed.), *2nd Pan-American Symposium on Landslides (II PSL/ 2a COBRAE)* (pp. 195–200), Rio de Janeiro.

Anbalagan, R., Srivastava, N. C. N., & Jain, V. (2000). Slope stability studies of Vyasi dma reservoir area, Garhwal Himalaya, U.P. India. In E. N. Bromhead, N. Dixon, & M.-L. Ibsen (Eds.), *Landslides in research, theory and practice, Proceedings of the 8th International Symposium on Landslides, 26–30 June 2000* (pp. 51–56). Cardiff: Thomas Telford.

Aniya, M. (1985). Landslide-susceptibility mapping in the Amahata River basin, Japan. *Annals of Argentina, Earth Surface Processes and Landforms, 29*, 255–266.

Ardizzone, F., Cardinali, M., Carrara, A., Guzzetti, F., & Reichenbach, P. (2002). Impact of mapping errors on the reliability of landslide hazard maps. *Natural Hazards and Earth Systems Sciences, 2*, 3–14.

Ayalew, L. (2000). Factors affecting slope stability in the Blue Nile Basin. In E. N. Bromhead, N. Dixon, & M.-L. Ibsen (Eds.), *Landslides in research, theory and practice, Proceedings of the 8th International Symposium on Landslides, 26–30 June 2000* (pp. 101–106). Cardiff: Thomas Telford.

Ayalew, L., & Yamagishi, H. (2005). The application of GIS based logistic regression for landslide susceptibility mapping in the Kakuda-Yahiko Mountains, Central Japan. *Geomorphology, 65*, 15–31.

Ayalew, L., Yamagishi, H., Marui, H., & Kanno, T. (2005). Landslides in Sado Island of Japan: Part II, GIS-based susceptibility mapping with comparisons of results from two methods and verifications. *Engineering Geology, 81*, 432–445.

Baeza, C., & Corominas, J. (2001). Assessment of shallow landslide susceptibility by Means of multivariate statistical techniques. *Earth Surface Processes and Landforms, 26*, 1251–1263.

Basu, S. R., & De, S. K. (2003). Causes and consequences of Landslides in the Darjiling and Sikkim Himalayas. *Geographia Polonica, 76*(2), 37–57.

Basu, S. R., & Ghatwar, L. (1988). *Landslide and soil erosion in the Gish basin of the Darjiling Himalayas and their bearing on North Bengal Floods. Studia Geomorpholigica Carpatho-Balcanica* (Vol. 22). Krakow.

Basu, S. R., & Maiti, R. K. (2001). Unscientific mining and degradation of slopes in the Darjeeling Himalayas, Changing environment. *Scenario of the Indian Subcontinent (Bd)*, 390–399.

Basu, S. R., & Sarkar, S. (1985). Some consideration on recent landslides at Tindharia and their control. *Indian Journal of Power and River Valley Development, 1985*, 190–194.

Barros, W. T., Amaral, C., & D'Orsi, R. N. (1991). Landslide susceptibility map of Rio de Janeiro, in D. H. Bell (Eds.), Landslides – *Proceedings of the Sixth International Symposium, 10–14 February 1992* (Rotterdam: A. A. Balkema), 869–871.

Basu, S. R., De, S. K., & Bera, B. (2009). Progress of research in landslide studies in India. In S. Singh, H. S. Sharma, S. K. De, H. S. Sharma, & V. S. Kale (Eds.), *Geomorphology in India* (pp. 171–198). Allahabad: Prayag Pustak Bhawan.

Bhandari, R. K. (1987). Slope stability in the fragile Himalaya and strategy for development, Ninth IGS Lecture. *Journal of the IGE, 17*(1), 1–78.

Bhattacharya, S. K. (1999). A constructive approach to landslides through susceptibility zoning and case study in the Rakti Basin of Eastern Himalaya, Transactions, Japanese Geomorphological.

Binaghi, E., Luzi, L., Madella, P., Pergalani, F., & Rampini, A. (1998). Slope Instability Zonation: A comparison between certainty factor and Fuzzy Dempster-Shafer approaches. *Natural Hazards, 17*(1), 77–97.

Brabb, E. E., Guzzetti, F., Mark, R., & Simpson, R. W. (1989). The extent of landsliding in Northern New Mexico and similar semi-arid regions. In D. M. Sadler & P. M. Morton (Eds.), *Landslides in a semi-arid environment* (Vol. 2, pp. 163–173). Publications of the Inland Geological Society.

Brardinoni, F., Slaymaker, O., & Hassan, M. A. (2003). Landslide inventory in a rugged forested watershed: A comparison between air-photo and field survey data. *Geomorphology, 54*, 179–195. https://doi.org/10.1016/S0169-555X(02)00355-0.

Carrara, A. (1989). *Landslide hazard mapping by statistical methods: A "black-box" model approach* (pp. 205–224). In Proceedings International Workshop on Natural Disasters in European-Mediterranean Countries, Perugia, June 27–July 1, 1988, CNR-ESNSF.

Carrara, A., Catalano, E., Sorriso Valvo, M., Reali, C., Merenda, L,. & Rizzo, V. (1977). *Bulletin of the International Association of Engineering Geology, 16*(1), 8–13.

Carrara, A., Cardinali, M., Guzzetti, F., & Reichenbach, P. (1995). *GIS technology in mapping landslide hazard. Geographical information systems in assessing natural hazards* (pp. 135–175). Dordrecht: Kluwer Academic Publishers.

Chakraborty, I., Ghosh, S., Bhattacharya, D., & Bora, A. (2011). Earthquake induced landslides in the Sikkim-Darjeeling Himalayas—An aftermath of the 18th September 2011 Sikkim earthquake. Geological Society of India.

Champati ray, P. K., Dimri, S., Lakhera, R. C., & Sati, S. (2007). Fuzzy-based method for landslide hazard assessment in active seismic zone of Himalaya. *Landslides, 4*, 101–111.

Chandler, R. J. (1972). Lias clay: weathering processes and their effect on shear strength. *Geotechnique, 15*, 403–431.

Chang, J. C., & Slaymaker, O. (2002). Frequency and spatial distribution of landslides in a mountainous drainage basin: Western Foothills, Taiwan. *Catena, 46*, 285–307.

Chen, C. L. (1997). *Debris-flow hazard mitigation: Mechanics, prediction and assessment*. San-Francisco: American Society of Civil Engineers.

Cheng, K. S., Wei, C., & Chang, S. C. (2004). Locating landslides using multi-temporal satellite images. *Advances in Space Research, 33*, 296–301.

Cheng, T. A., Lateh, H., & Peng, K. S. (2008). *Intelligence explanation system on landslide dissemination: A case study in Malaysia* (pp. 330–333). In Proceedings of the first world landslide forum report: Implementing the 2006 Tokyo action plan on the international program on landslides (IPL).

Cherkez, Y., Kozlova, V., Shmourakto, V., Kharitonov, V., & Karavan, A. (2000). Landslide in the North-Western Black sea region. In E. N. Bromhead, N. Dixon, & M.-L. Ibsen (Eds.), *Landslides in research, theory and practice, Proceedings of the 8th International Symposium on Landslides, 26–30 June 2000* (pp. 251–254). Cardiff: Thomas Telford.

Choi, J., Oh, H. J., Lee, H., Lee, C., & Lee, S. (2011). Combining landslide susceptibility maps obtained from frequency ratio, logistic regression, and artificial neural network models using aster images and GIS. *Engineering Geology, 124*, 12–23.

Chung, C. F., & Fabbri, A. G. (1999). Probabilistic prediction models for landslide hazard mapping. *Photogrammetric Engineering and Remote Sensing, 65*(12), 1389–1399.

Chung, C.-J. F., Fabbri, A. G., & van Westen, C. J. (1995). Multivariate regression analysis for landslide hazard zonation. In A. Carrara & F. Guzzetti (Eds.), *Geographical information systems*.

Claessens, L., Verburg, P. H., Schoorl, J. M., & Veldkamp, A. (2006). Contribution of topographically based landslide hazard modelling to the analysis of the spatial distribution and ecology of kauri (Agathis australis). *Landscape Ecology, 21*, 63–76.

Committee on the Review of the National Landslide Hazards Mitigation Strategy. (2004). *Partnerships for reducing landslide risk, assessment of the National Landslide Hazards*

Mitigation Strategy, Board on Earth Sciences and Resources, Division on Earth and Life Studies, The National Academic Press, Washington, D.C. (p. 143).

Coussot, P., Laigle, D., Arattano, M., Deganutti, A. M., & Marchi, L. (1998). Direct determination of rheological characteristics of debris-flow. *Journal of Hydraulic Engineering, 124*, 865–868.

Crovelli, R. A. (2000). Probability Models for estimation of number and costs of landslides, United States Geological Survey Open –File Report, 00–249.

Cuesta, M. J. D., Sánchez, M. J., & García, A. R. (1999). Press archives as temporal records of landslides in the North of Spain: Relationships between rainfall and instability slope events. *Geomorphology, 30*, 125–132.

Dai, F. C., & Lee, C. F. (2001). Terrain-based mapping of landslide susceptibility using a geographical information system: A case study. *Canadian Geotechnical Journal, 38*, 911–923.

Dai, F. C., & Lee, C. F. (2002). Landslides on natural terrain—Physical characteristics and susceptibility mapping in Hong Kong. *Mountain Research and Development, 22*, 40–47.

Dai, F. C., Lee, C. F., & Ngai, Y. Y. (2002). Landslide risk assessment and management: An overview. *Engineering Geology, 64*, 65–87.

Das, I., Kumar, G., Stein, A., Bagchi, A., & Dadhwal, V. K. (2011). Stochastic landslide vulnerability modeling in space and time in a part of the northern Himalayas, India. *Environmental Monitoring and Assessment, 178*, 25–37.

De Jaeger, C. (2000). Influence on landsliding and slope development of the particular environment of the Dead Sea region: A case study for the Wadi Mujib Canyon (Jordan). In E. N. Bromhead, N. Dixon, & M.-L. Ibsen (Eds.), *Landslides in research, theory and practice, Proceedings of the 8th International Symposium on Landslides, 26–30 June 2000* (pp. 421–426). Cardiff: Thomas Telford.

Dehn, M., Burger, G., Buma, J., & Gasparetto, P. (2000). Impact of climate change on slope stability using expanded downscaling. *Engineering Geology, 55*, 193–204.

Devkota, K. C., Regmi, A. D., Pourghasemi, H. R., Yoshida, K., Pradhan, B., Ryu, I. C., Dhital, M. R., & Althuwaynee, O. F. (2013). Landslide susceptibility mapping using certainty factor index of entropy and logistic regression models in GIS and their comparison at Mugling-Narayanghat road section in Nepal Himalaya. *Natural Hazards, 65*, 135–165.

Dhakal, A. S., Amada, T., & Aniya, M. (2000). Landslide hazard mapping and its evaluation using GIS: An investigation of sampling schemes for a grid-cell based quantitative method. *Photogrammetric Engineering & Remote Sensing, 66*, 981–989.

Dikau, R., Brunsden, D., Schrott, L., & Ibsen, M. L. (Eds.). (1996). *Landslide recognition: Identification, movement, and causes* (p. 210). New York: Wiley.

Donati, L., & Turrini, M. C. (2002). An objective method to rank the importance of the factors predisposing to landslides with the GIS methodology: Application to an area of the Apennines (Valnerina; Perugia, Italy). *Engineering Geology, 63*, 277–289.

Doratti, M., McColl, C., & Tweeddale, C. (2002). *Landslide prediction study BCIT, Geographical Information Systems advanced diploma*. giswww1bcitca/giscentre/projects/projects2002/prj_09_final_reportpdf.

Erickson, G. E., Ramirez, C. F., Concha, J. F., Tisnado, G. M., & Urquidi, F. B. (1989). Landslide hazards in the central and southern Andes. In E. E. Brabb & B. L. Harrod (Eds.), *Landslides: extent and economic significance* (pp. 111–118). Rotterdam: A.A. Balkema.

Evans, S. G., & DeGraff, J. V. (Eds.). (2002). *Catastrophic landslides, reviews in engineering geology*. Boulder: Geological Society of America.

Evans, S. G., & Hungr, O. (1993). The assessment of rock fall hazards at the base of talus slopes. *Canadian Geotechnical Journal, 30*, 620–636.

Fabbri, A. G., Chung, C.-J., Napolitano, P., Remondo, J., & Zêzere, J. L. (2002). Prediction rate functions of landslide susceptibility applied in the Iberian Peninsula, In C. A. Brebbia (Ed.), *Third International Conference on Risk Analysis, 19–21 June 2002*, Sintra, Portugal, pp. 703–718.

Fernandez-Steeger, T. M., Rohn, J., & Czurda, K. (2002). Identification of landslide areas with neural networks for hazard analysis. Landslides. In: *Proceedings of the first European conference on landslides*, (pp. 163–168). Prague, Czech Republic; Rotterdam, Balkema.

Garfi, G., & Bruno, D. E. (2007). Fan morphodynamics and slope instability in the Mucone River basin (Sila Massif southern Italy): Significance of weathering and role of land use changes. *Catena, 50*, 181–196.

Ghosh, S., Van Westen, C. J., Carranza, E., & Jetten, V. (2009). Generation of event-based landslide inventory maps in a data-scarce environment; case study around Kurseong, Darjiling District, West Bengal, India. In P. Malet, A. Remaitre, & T. Bogaard (Eds.), *Landslide processes: From geomorphologic mapping to dynamic modeling: Proceedings of the landslide processes* (pp. 37–44). Strasbourg: European Centre on Geomorphological Hazards (CERG).

Giannecchini, R., Duccio Naldini, D., Avanzi, G. D., & Puccinelli, A. (2007). Modelling of the initiation of rainfall-induced debris flows in the Cardoso basin (Apuan Alps, Italy). *Quaternary International, 171*, 108–117.

Glade, T. (1997). *The temporal and spatial occurrence of rainstorm-triggered landslide events in New Zealand.* PhD Thesis, Victoria University of Wellington, Wellington.

Glade, T. (2001). Landslide hazard assessment and historical landslide data—An inseparable couple? In T. Glade, F. Frances, & P. Albini (Eds.), *The use of historical data in natural hazard assessments* (Vol. 7, pp. 153–168). Dordrecht: Kluwer Academic Publishers.

Gorsevski, P. V., Jankowski, P., & Gessler, P. E. (2006). An heuristic approach for mapping landslide hazard by integrating fuzzy logic with analytic hierarchy process. *Control and Cybernetics, 35*(1), 121–146.

Gupta, S. K. (2000). 1897 Great Assam earthquake-generated landslides: Distribution, pattern and correlation with landslide potentiality of Northeastern India. In E. N. Bromhead, N. Dixon, & M.-L. Ibsen (Eds.), *Landslides in research, theory and practice, Proceedings of the 8th International Symposium on Landslides, 26–30 June 2000* (Vol. 2, pp. 677–682). Cardiff: Thomas Telford.

Guzzetti, F., Carrara, A., Cardinali, M., & Reichenbach, P. (1999). Landslide hazard evaluation: A review of current techniques and their application in a multi-scale study central Italy. *Geomorphology, 31*(1–4), 181–216.

Guzzetti, F., Crosta, G., Detti, R., & Agliardi, F. (2002). STONE: A computer programme for the three-dimensional simulation of rock-falls. *Computers and Geosciences, 28*, 1079–1093.

Guzzetti, F., Reichenbach, P., Cardinali, M., Galli, M., & Ardizzone, F. (2005). Probabilistic Landslide hazard assessment at the basin scale. *Geomorphology, 72*, 272–299.

Guzzetti, F., Galli, M., Reichenbach, P., Ardizzone, F., & Cardinali, M. (2006). Landslide hazard assessment in the Collazzone area, Umbria, Central Italy. *Natural Hazards and Earth System Science 6*(1):115–131.

Hayne, M. C., & Gordon, D. (2001). Regional landslide hazard estimation, a GIS/decision tree analysis: Southeast Queensland, Australia. In K. K. S. Ho & K. S. Li (Eds.), *Geotechnical Engineering–Meeting Society's Needs, Proceedings of the 14th Southeast Asian Geotechnical Conference,10–14 December 2001* (pp. 115–121). Hong Kong: A.A. Balkema.

Heckerman, D. E. (1986). Probabilistic interpretation of MYCIN's certainty factors. In *Reading in uncertain reasoning* (pp. 298–312). San Francisco: Morgan Kaufmann Publishers Inc.

Highland, L. M., & Bobrowsky, P. (2008). *The landslide handbook—A guide to understanding landslides* (p. 129). Reston: U.S. Geological Survey Circular 1325.

Holm, K., Bovis, M., & Jakob, M. (2004). The landslide response of alpine basins to post-Lille Ice Age glacial thinning and retreat in southwestern British Columbia. *Geomorphology, 57*, 201–216 Honolulu, HI. Gen. Tech. Rep. WO-26. Washington, DC: U.S. Department of Agriculture, Forest Service: 401–420.

Hungr, O. (2005). Classification and terminology. In M. Jakob & O. Hungr (Eds.), *Debris flow hazard and related phenomena*. Heidelberg: Springer.

Hutchinson, J. N. (1995). *Landslide hazard assessment* (Vol. 1, pp. 1805–1842). In Proc VI Int Symp on Landslides, Christchurch.

Hutchinson, J. N., & Bromhead, E. N. (2002). Isle of Wight landslides. In R. G. McInnes & J. Jakeways (Eds.), *Instability planning and management* (pp. 3–72). Isle of Wight: Thomas Telford.

Jibson, R. (2000). A method for producing digital probabilistic seismic landslide hazard maps. *Engineering Geology, 58*, 271–289.

Jibson, R. W., Harp, E. L., & Michael, J. A. (2000). A method for producing digital probabilistic seismic landslide hazard maps. *Engineering Geology, 58*, 271–289.

Jonasson, C., Nyberg, R., & Rapp, A. (1997). Dating of rapid mass movements in Scandinavia: Talus rockfalls, large rockslides, debris flows and slush avalanches. In J. A. Matthews, D. Brunsden, B. Frenzel, B. Gläser, & M. M. Weiß (Eds.), *Rapid mass movement as a source of climatic evidence for the Holocene* (Vol. 12, pp. 267–282). Stuttgart, Jena, Lübeck and Ulm: Gustav Fischer Verlag.

Jones, C. L., Higgins, J. D., & Andrew, R. D. (2000). *Colorado rockfall simulation programme*. Denver: Colorado Department of Transportation.

Kamai, T., Kobayashi, Y., Jinbo, C., & Shuzui, H. (2000). *Earthquake risk assessments of fill-slope instability in urban residential areas in Japan*. In E. N. Bromhead, N. Dixon, & M.-L. Ibsen (Eds.).

Kanungo, D. P., Arora, M. K., Sarkar, B. S., & Gupta, R. P. (2006). A comparative study of conventional ANN black box fuzzy and combined neural and fuzzy weighting procedures for landslide susceptibility zonation in Darjeeling Himalayas. *Engineering Geology, 85*, 347–366.

Kayastha, P., Dhital, M. R., & De Smedt, F. (2013). Application of the analytical hierarchy process (AHP) for landslide susceptibility mapping: A case study from the Tinau watershed, west Nepal. *Computers & Geosciences, 52*, 398–408.

Kim, W. Y., Chae, B. G., Kim, K. S., & Cho, Y. C. (2001). Approach to quantitative prediction of landslides on natural mountain slopes, Korea. In K. K. S. Ho & K. S. Li (Eds.), *Geotechnical Engineering—Meeting Society's Needs, Proceedings of the 14th Southeast Asian Geotechnical Conference, 10–14 December 2001* (pp. 795–799). Hong Kong: A.A. Balkema.

Korup, O. (2003). *Landslide-induced river disruption—Geomorphic imprints and scaling effects in alpine catchments of South Westland and Fiordland, New Zealand*. Unpubl. PhD thesis. Victoria, University of Wellington, p. 314.

Larsen, M. C., & Torres-Sanchez, A. J. (1998). The frequency and distribution of recent landslides.

Lee, E. M., & Clark, A. R. (2000). The use of archive records in landslide risk assessment: historical landslide events on the Scarborough coast, UK. In E. N. Bromhead, N. Dixon, & M.-L. Ibsen (Eds.), *Landslides in research, theory and practice, Proceedings of the 8th International Symposium on Landslides, 26–30 June 2000* (pp. 905–910). Cardiff: Thomas Telford.

Lee, S., & Pradhan, B. (2006). Probabilistic landslide hazard and risk mapping on Penang Island, Malaysia. *Journal of Earth System Science, 115*, 661–672. https://doi.org/10.1007/s12040-006-0004-0.

Lee, S., & Pradhan, B. (2007). Landslide hazard mapping at Selangor, Malaysia using frequency ratio and logistic regression models. *Landslides, 4*(1), 33–41.

Lee, S., & Sambath, T. (2006). Landslide susceptibility mapping in the Damrei Romel area Cambodia using frequency ratio and logistic regression models. *Environmental Geology, 50*, 847–855.

Lee, C. F., Ye, H., Yeung, M. R., Shan, X., & Chen, G. (2001). AIGIS-based methodology for natural terrain landslide susceptibility mapping in Hong Kong. *Episodes, 24*(3), 150–159.

Lee, S., Chwae, U., & Min, K. D. (2002). Landslide susceptibility mapping by correlation between topography and geological structure: The Janghung area, Korea. *Geomorphology, 46*, 149–162.

Lee, S., Ryu, J. H., Min, K. D., & Won, J. S. (2003). Landslide susceptibility analysis using GIS and artificial neural network. *Earth Surface Processes and Landforms, 27*, 1361–1376.

Lee, S., Choi, J., & Woo, I. (2004). The effect of spatial resolution on the accuracy of landslide susceptibility mapping: A case study in Boun, Korea. *Geosciences Journal, 8*, 51–60.

Lee, S., Ryu, J. H., & Kim, I. S. (2007). Landslide susceptibility analysis and its verification using likelihood ratio logistic regression and artificial neural network models: Case study of Youngin Korea. *Landslides, 4*, 327–338.

Li, C., Ma, T., Sun, L., Li, W., & Zheng, A. (2011). Application and verification of fractal approach to landslide susceptibility mapping. *Natural Hazards*. https://doi.org/101007/s11069-011-9804-x.

Lorente, A., Garcia-Ruiz, J. M., Begueria, S., & Arnaez, J. (2002). Factors explaining the spatial distribution of hill slope debris flow: A case study in the flysch sector of the central Spanish Pyrenees. *Mountain Research and Development, 22*, 32–39.

Maharaj, R. J. (1993). Landslide processes and landslide susceptibility analysis from an upland watershed: A case study from St. Andrew, Jamaica, West Indies. *Engineering Geology, 34*, 53–79.

Maiti, R. (2007). *Identification of potential slope failure zones of shiv-khola watershed; Darjiling Himalaya, through critical analysis of slope instability—A step towards rational and scientific management of land, soil and water*. UGC sponsored minor research project [F.31-210/2005 (31.03.2007)].

Mancini, F., Ceppi, C., & Ritrovato, G. (2010). GIS and statistical analysis for landslide susceptibility mapping in the Daunia area, Italy. *Natural Hazards and Earth System Sciences, 10*, 1851–1864. https://doi.org/10.5194/nhess-10-1851-2010.

Mandal, S., & Maiti, R. (2013). Integrating the Analytical Hierarchy Process (AHP) and the frequency ratio (FR) model in landslide susceptibility mapping of Shiv-khola watershed, Darjeeling Himalaya. *International Journal of Disaster Risk Science, 4*(4), 200–212.

Mandal, B., & Mandal, S. (2016). Assessment of mountain slope instability in the Lish River basin of Eastern Darjeeling Himalaya using frequency ratio model (FRM). *Modeling Earth Systems and Environment, 2*, 121.

Mansouri Daneshvar, M. R. (2014). Landslide susceptibility zonation using analytical hierarchy process and GIS for the Bojnurd region, northeast of Iran. *Landslides, 11*, 1079–1091.

Mark, R. K., & Ellen, S. D. (1995). Statistical and Simulation Models for mapping Debris-flow hazard. In A. Carrara & F. Guzzetti (Eds.), *Geographical information systems in assessing natural hazards* (Vol. 5, pp. 93–106). Dordrecht: Kluwer Academic Publishers.

Martinez, J. M., Avila, G., Agudelo, A., Schuster, R. L., Casadevall, T. J., & Scott, K. M. (1995). Landslides and debris flows triggered by the 6 June 1994 Paez earthquake, southwestern Columbia. *Landslide News, 9*, 13–15.

Meng, X., Derbyshire, E., & Du, Y. (2000). Landslide hazard in the eastern part of Gansu Province, China. *Zeitschrift der Geologischen Gesellschaft, 151*, 31–47.

Moeyersons, J., Tre'fois, P. H., Lavreau, J., Alimasi, D., Badriyo, I., Mitima, B., Mundala, M., Munganga, D. O., & Nahimana, L. (2004). A geomorphological assessment of landslide origin at Bukavu, Democratic Republic of the Congo. *Engineering Geology, 72*, 73–87.

Mondal, S., & Maiti, R. (2012). Landslide susceptibility analysis of Shiv-Khola watershed, Darjiling: A remote sensing & GIS based Analytical Hierarchy Process (AHP). *Journal of the Indian Society of Remote Sensing, 40*, 483–496.

Mondal, S., & Mandal, S. (2017). Application of frequency ratio (FR) model in spatial prediction of landslides in the Balason River Basin, Darjeeling Himalaya. *Spatial Information Research, 25*(3), 337–350. https://doi.org/10.1007/s41324-017-0101-y.

Monma, K., Kojima, S., & Kobayashi, T. (2000). Rock slope monitoring and rock fall prediction. *Landslide News, 13*, 33–35.

Moreiras, S. (2004). Landslide incidence zonation in the Rio Mendoza valley, Mendoza Province, morphometry and typology in two zones, Calabria, Italy. *Bulletin of the International Association of Engineering Geology, 16*, 8–13.

Moser, M. (2002). Geotehnical aspects of landslides in the Alps. In J. Rybar, J. Stemberk, & P. Wagner (Eds.), *Landslides, 24–26 June 2002, Prague, Czech Republic* (pp. 23–44). Rotterdam: A.A.Balkema.

Mousavi, S. Z., Kavian, A., Soleimani, K., Mousavi, S. R., & Shirzadi, A. (2011). GIS-based spatial prediction of landslide susceptibility using logistic regression model. *Geomatics Natural Hazards and Risk, 2*, 33–50.

Naranjo, J. L., van Westen, C. J., & Soeters, R. (1994). Evaluating the use of training areas in bivariate statistical landslide hazard analysis—A case study in Columbia. *ITC Journal, 3*.

NDMA. (2009). Management of Landslides and Snow Avalanches, National Disaster Management Authority (NDMA) Government of India New Delhi, 144.

Page, M. J., Trustrum, N. A., & Dymond, J. R. (1994). Sediment budget to assess the geomorphic effect of a cyclonic storm, New Zealand. *Geomorphology, 9*, 169–188.

Panizza, M., Pasuto, A., Silvano, S., & Soldati, M. (1996). Temporal occurrence and activity of Perugia, 27 June–1 July 1988 (CNR-ESNSF), 205-224.

Pistocchi, A., Luzi, L., & Napolitano, P. (2002). The use of predictive modeling techniques for optimal exploitation of spatial databases: A case study in landslide hazard mapping with expert system-like methods. *Environmental Geology, 41*, 765–775.

Pradhan, B. (2010). Remote sensing and GIS-based landslide hazard analysis and cross validation using multivariate logistic regression model on three test areas in Malaysia. *Advances in Space Research, 45*, 1244–1256.

Pradhan, B., & Lee, S. (2009). Delineation of landslide hazard areas using frequency ratio logistic regression and artificial neural network model at Penang Island Malaysia. *Environmental Earth Science, 60*, 1037–1054.

Rautelal, P., & Lakheraza, R. C. (2000). Landslide risk analysis between Giri and Tons Rivers in Himachal Himalaya, India. *International Journal of Applied Earth Observation and Geoinformation, 2*, 153–160.

Rickenmann, D., & Chen, C. L. (2003). *Debris-flow hazard mitigation: Mechanics, prediction and assessment, 10–12 September 2003, Davos, Switzerland* (p. 1335). Rotterdam: Millpress.

Saaty, T. L. (1980). *The analytical hierarchy process* (p. 350). New York: McGraw Hill.

Saaty, T. L. (1990). *The analytical hierarchy process: planning, priority setting, resource allocation* (1st ed.p. 502). Pittsburgh: RWS Publication.

Saaty, T. L. (1994). *Fundamentals of decision making and priority theory with analytic hierarchy process* (1st ed.p. 527). Pittsburgh: RWS Publication.

Sabto, M. (1991). *Probabilistic modeling applied to landslides in Central Colombia using GIS procedures*. Unpublished MSc thesis, ITC, Enschede, Netherlands, p. 26.

Saha, A. K., Gupta, R. P., Sarkar, I., Arora, M. K., & Csaplovics, E. (2005). An approach for GIS based statistical landslide susceptibility zonation with a case study in the Himalayas. *Landslides, 2*, 61–69.

Sarkar, S. (1999a). Landslides in Darjeeling Himalayas, India. *Transactions, Japanese Geomorphological Union, 20*, 299–315.

Sarkar, S. (1999b). Landslides in Darjiling Himalayas. *Transactions, Japanese Geomorphological Union, 20*(3), 299–315.

Sarkar, S., Kanungo, D., Patra, A., & Kumar, P. (2006). *Disaster mitigation of debris flows, slope failures and landslides, GIS based landslide susceptibility mapping—A case study in Indian Himalaya* (pp. 617–624). Tokyo: Universal Academy Press.

Schmidt, K. H., & Beyer, I. (2001). Factors controlling mass movement susceptibility on the Wellenkalk-scarp in Hesse and Thuringia. *Zeitschrift für Geomorphologie, Supplementband, 125*, 43–63.

Schmidt, K. H., & Beyer, I. (2003). High magnitude landslide-events on a limestone-scarp in central Germany—Morphometric characteristics and climatic controls. *Geomorphology, 49*, 323–342.

Schmidt, K. M., Roering, J. J., Stock, J. D., Dietrich, W. E., Montgomery, D. R. & Schaub, T. (2001). Root cohesion variability and shallow landslide susceptibility in the Oregon Coast Range. *Canadian Geotechnical Journal 38*(1): 995–1024.

Schoeneich, P., & Bouzou, I. (1996). Landslides in Niger (West Africa). In Senneset (Ed.), *Landslides* (pp. 1967–1972). Rotterdam: Balkema.

Sharpe, C. F. S. (1938). *Landslides and related phenomena*. New York: Columbia University Press.

Shorthliffe, E. H., & Buchanan, G. G. (1975). A model of inexact reasoning in medicine. *Mathematical Biosciences, 23*, 351–379.

Shroder, J. F., & Bishop, M. P. (1998). Mass Movement in the Himalaya: New insights and research directions. *Geomorphology, 26*, 13–35.

Soeters, R., & van Westen, C. J. (1996). Slope instability, recognition, analysis and zonation. In A. K. Turner & R. L. Schuster (Eds.), *Landslide: investigations and mitigation. Special Report*

247. Transportation Research Board. National Research Council (pp. 129–177). Washington, D.C.: National Academy Press.

Song, Y., Gong, J., Gao, S., Wang, D., Cui, T., Li, Y., & Wei, B. (2012). Susceptibility assessment of earthquake induced landslides using Bayesian network: A case study in Beichuan China. *Computational Geosciences, 42*, 189–199.

Spang, R. M., & Sonser, T. (1995). Optimized rockfall protection by 'Rockfall'. In T. Fuji (Ed.), *8, International Congress on Rock Mechanics (Tokyo, Japan)* (pp. 1233–1242).

Starkel, L. (1972). The role of catastrophic rainfall in the shaping of the relief of the Lower Himalaya (Darjeeling Hills). *Geographia Polonica, 21*, 103–147.

Starkel, L., & Basu, S. R. (2000). *Rains, landslides and floods in the Darjiling Himalayas*. New Delhi: Indian Science Academy.

Stead, D. S., Eberhardt, E., Coggan, J., & Bemko, B. (2001). Advanced numerical techniques in rock slope stability analysis—Application and limitations. In M. Kuhne, H. H. Einstein, E. Krauter, H. Klapperich, & R. Pottler (Eds.), *Landslides-causes, impacts and counter measures, 17–21 June 2001*. Davos: Verlag Gluckauf Essen.

Swanson, F. J., & Dyrness, C. T. (1975). Impact of clearcutting and road construction on soil erosion by Landslides in the Western Cascade Range, Oregon. *Geology, 3*, 393–396.

Tang, C., & Grunert, J. (1999). Inventory of landslides triggered by the 1996 Lijiang earthquake, Yunnan Province, China. *Transactions, Japanese Geomorphological Union, 20*, 335–349.

Tavares, A. O., & Soares, A. F. (2002). Instability relevance on land use planning in Coimbra municipality (Portugal). In R. G. McInnes & J. Jakeways (Eds.), *Instability planning and management* (pp. 177–184). Isle of Wight: Thomas Telford.

Tazik, E., Jahantab, Z., Bakhtiari, M., Rezaei, A., & Alavipanah, S. K. (2014). *Landslide susceptibility mapping by combining the three methods Fuzzy Logic, Frequency Ratio and Analytical Hierarchy Process in Dozain basin, The International Archives of the Photogrammetry, Remote Sensing and Spatial Information Sciences, Volume XL-2/W3, 2014*. In 1st ISPRS International Conference on Geospatial Information Research, 15–17 November 2014, Iran: Tehran.

The Hindu. (2015). *30 killed in Darjeeling landslides*. New Delhi: The Hindu.

Tibaldi, A., Ferrari, L., & Pasquare, G. (1995). Landslides triggered by earthquakes and their relations with faults and mountain slope geometry—An example from Ecuador. *Geomorphology, 11*, 215–226.

Twidale, C. R. (2000). The Lochiel landslip, a mass movement developing in 1974 but originating 600–700 million years earlier. In E. N. Bromhead, N. Dixon, & M.-L. Ibsen (Eds.), *Landslides in union* (Vol. 20, pp. 317–333).

Uromeihy, A. (2000). Use of landslide hazard zonation map in the evaluation of slope instability in the Jiroft-dam watershed. In E. N. Bromhead, N. Dixon, & M.-L. Ibsen (Eds.), *Landslides in research, theory and practice, Proceedings of the 8th International Symposium on Landslides, 26–30 June 2000* (pp. 1501–1508). Cardiff: Thomas Telford.

van Westen, C. J., & Getahun, L. F. (2003). Analyzing the evolution of the Tessina landslide using aerial photographs and digital elevation models. *Geomorphology, 54*, 77–89.

van Westen, C. J., Rengers, N., Terlien, M. T. J., & Soeters, R. (1997). Prediction of the occurrence of slope instability phenomena through GIS-based hazard zonation. *Geologische Rundschau, 86*(2), 404–414.

van Westen, C. J., Van Asch, T. W. J., & Soeters, R. (2006). Landslide zonation and risk hazard—Why is it still so difficult? *Bulletin Engineering Geology and Environment, 65*, 167–184.

Varnes, D. J. (1978). Slope movement types and processes. In: Schuster, R. L, & Krizek, R. J. (Eds.), *Landslides, analysis and control, special report 176: Transportation research board, National Academy of Sciences*, (pp. 11–33). Washington, DC.

Varnes, D. J. (1984). Landslide Hazard Zonation: A review of principles and practice, Commission on Landslides of the IAEG, UNESCO. *Natural Hazards, 3*, 61.

Wan, S. (2012) Entropy-based particle swarm optimization with clustering analysis on landslide susceptibility mapping. *Environmental Earth Sciences*. https://doi.org/101007/s12665-012-1832-7.

Wieczorek, G. F., & Naeser, N. D. (2000). *Proceeding of the Second International Conference on Debris-flow hazard mitigation, 16-18 August 2000 Debris-flow hazard mitigation: Mechanics, prediction and Assessment, Taipei, Taiwan* (p. 608). Rotterdam: Balkema.

Wu, C. H., & Chen, S. C. (2009). Determining landslide susceptibility in Central Taiwan from rainfall and six site factors using the analytical hierarchy process method. *Geomorphology, 11*, 190–204.

Xie, M., Esaki, T., & Cai, M. (2004). A GIS-based method for locating the critical 3D slip surface in a slope. *Computers and Geotechnics, 31*, 267–277.

Xu, C., Dai, F., Xu, X., & Lee, Y. H. (2012). GIS-based support vector machine modeling of earthquake-triggered landslide susceptibility in the Jianjiang River watershed China. *Geomorphology, 145–146*, 70–80.

Yamagami, T., Jiang, J. C., & Yokino, K. (2001). An identification of DEM parameters for rock fall analysis. In K. K. S. Ho & K. S. Li (Eds.), *Geotechnical Engineering-Meeting Society's Needs, Proceedings of the 14th South East Asian Geotechnical Conference, 10–14 December 2001* (pp. 953–958). Hong Kong: A.A. Balkema.

Yu, F. C., Che, T. C., Lin, K. L., Chen, C. Y., & Yu, W. H. (2006). Landslides and rainfall characteristics analysis in Taipei city during the Typhoon Nari event. *Natural Hazards, 37*, 153–167.

Zezere, J. L. (2002). Landslide susceptibility assessment considering landslide typology, a case study in the area north of Lisbon (Portugal). *Natural Hazards and Earth System Sciences, 2*, 73–82.

Zhou, C. H., Lee, C. F., Li, J., & Xu, Z. W. (2002). On the spatial relationship between landslides and causative factors on Lantau Island, Hong Kong. *Geomorphology, 43*, 197–207.

Zorn, M., & Komac, B. (2004). Deterministic modeling of landslide and rockfall risk. *Acta Geographica Slovenica, 44*, 53–100.

Chapter 2
Geomorphic, Geo-tectonic, and Hydrologic Attributes and Landslide Probability

2.1 Introduction

The study of various geomorphic, geo-tectonic, and geo-hydrologic factors and their proper understanding give an account on the geomorphological and evolutionary character of the landform. The changes and evolution of landscape could be studied by studying all the factors which will help to restore proper management technique for proper utilization of land and soil. In the present study, the landslide influencing factors such as slope angle, slope aspect, slope curvature, geomorphology, geology, soil, drainage, lineament, elevation, topographic wetness index, stream power index, land use, and land cover of Darjeeling Himalaya were generated and their relationship with the landslide inventory was made on GIS platform. Landslide susceptibility of each class of the landslide conditioning factors were assessed estimating the frequency ratio value of individual class after Lee and Sambath (2006) and Lee and Pradhan (2007). Beven and Kirkby (1979), and Montgomery and Dietrich (1994) and Dietrich et al. (1995, 2001) introduced so many process-based model in relation to landslides. Darjeeling Himalaya represents diversified topographic phenomena which not only influences flow properties but also slope material properties and landslide as well. The assessment of various geomorphic, geo-tectonic, and geo-hydrologic properties were assessed in detail which helped to perform various statistical models in monitoring and assessing landslides susceptibility in Darjeeling Himalaya, West Bengal. All the prepared attribute layers were converted into raster value domain on GIS environment and were integrated with the landslide inventory to identify the relative importance with regard to landslide potentiality of all the attribute class. The landslide inventory map is one of the significant data layers which helped to assess the probability of landslide, landslide susceptibility, and accuracy level. Landslide inventory map identified landslide location, landslide types, and spatial and temporal pattern of landslide in a region. This distribution pattern of landslide phenomena were obtained through intensive field investigation, historical landslide records, satellite images, and aerial

S. Mandal, S. Mondal, *Statistical Approaches for Landslide Susceptibility Assessment and Prediction*, https://doi.org/10.1007/978-3-319-93897-4_2

photograph. Landslide inventory map is "the simplest form of landslide information which records the location, date of occurrence, type that have also left some identifiable imprints in the area" (Cruden 1991). Guzzetti et al. (2005), Colombo et al. (2005), and Galli et al. (2008) prepared landslide inventory map in a scientific and systematic way and used it to prepare landslide susceptibility map using suitable statistical modelling with the help of GIS. To find out the relative significance of all the landslide conditioning factors and their classes, FR ratio values for each class was derived after Lee and Sambath (2006) and Lee and Pradhan (2007). The estimation of FR value considered total number of pixels affected by landslide, total number of pixels prevalent in the study area, class wise number of pixels and class wise number of pixels affected by landslides. The frequency ratio value indicates the density and probability of landslide in each class and accordingly class weight values were assigned to perform statistical model as well as to prepare the landslide susceptibility map of Darjeeling Himalaya.

It is true that geomorphic, geo-tectonic, and geo-hydrologic parameters change geomorphic threshold of the mountain slope materials and influence the stability character of the landform. Geomorphic thresholds, i.e. critical slope angle, critical slope height, and critical rainfall, are determined based on the estimated slope angle, soil moisture condition, depth of the soil, angle of internal friction, transmissivity, permeability, pore-water pressure, upslope contributing area, cohesion, etc. for a specific location of the mountain slope. All these parameters help the slope surface material to move downward by increasing driving force under the influence of gravity. The factors which are responsible for landslides or down slope movement of the slope materials can be categorized into four such as preconditioning factors/predisposing factors, preparatory factors, triggering factors, and sustaining factors.

Preparatory factors are those which are being facilitated by some preconditions and which are responsible for mass movement and landslide. Reduction of the strength of rocks by weathering processes, changes in the climatic (rainfall) conditions, tectonic upliftment, and human intervention-induced modification of slope angle, slope height, increase in the pressure over the slope materials with the development of concrete structures are included in the preparatory factors of landslides. Chandler (1972), Dehn et al. (2000), Shroder and Bishop (1998), Preston (2000), Schmidt et al. (2001), and Rybar (1997) analysed the significance of all the preparatory factors and their role in the reduction of the margin of stability in the mountain slope. *Triggering factors* changes the slope materials from stable to unstable conditions. *High intensity of rainfall* within a short time span may saturate the slope materials and reduce cohesion and shearing strength as well as landslides. *Tectonic movement* and associated *tremor* and *shaking* change the stability situation and promote actively unstable state of the mountain slope. Human intervention caused due to large-scale deforestation exposed the slope to atmospheric and weathering processes which increases surface run-off, soil erosion, and landslip. *Undercutting of the slope* for the construction of roads and development of the tea garden area steepened the slope and lost the basal support of the upslope materials which promote driving force and invite landslide in the valley side slopes. Crozier (1989) and Glade (1998) suggested that the spatial distribution of landslide susceptibility can be assessed by estimating the threshold values of different locations of the concerned study area. The estimated threshold values could be used for forecasting the landslide phenomena over the region (Crozier 1999).

Darjeeling Himalaya consists of two distinct tracts: the terai immediately beneath the hills and the ridges and deep valleys of the lower Himalayas. The terai region is traversed by numerous rivers and streams rushing down from the hilly areas of Darjeeling Himalaya. The region is now cleared and converted into tea garden area and settled villages. Behind the terai region, lofty mountain ranges are being found containing a series of ridges and valleys. In 1830, the mountain areas were covered with clothed forest from the divide summit valley bottoms. Darjeeling Himalaya contains a major mountain range viz. Singalila range commanding Nepal, Sikkim, and Darjeeling. Other ranges are Ghum range which meets the Singalila range and where elevation ranges from 7000 to 7900 ft; the Senchal-Mahaldiram range directed towards south and reaching an elevation of 8000–8600 ft, and this range consists of Tiger Hill; the Takdah range branches off from the main Senchal ridge below the Tiger Hill where elevation ranges from 7300 to 680 ft; and the Darjeeling Jalpaiguri range which extends northwards from Ghum at an elevation of 7886 ft and descends to 7002 ft in Darjeeling (Chaurasta). Due to such topographic diversity and existence of several mountain ranges, Darjeeling Himalayan mountain region is characterized by active geomorphological processes, i.e. fluvial processes, weathering, mass wasting, and landslides. All these activities are frequently jeopardizing in the Himalayan mountain slope making the slope segment more vulnerable to landslides.

2.2 Tectonic Settings of Darjeeling Himalaya

The Himalaya is a young fold mountain chain originated as a result of collision between the Indian plate and the Eurasian plate around 48 Ma before present (Le Fort 1975). The lesser Himalaya is tectonically confined between the Main Central Thrust (MCT) and the Main Boundary Thrust (MBT) and where the MBT is constituted due to deformed Proterozoic rock sequence dating back to 1.8 Ga (Kumar 1997; Yin et al. 2010). Geo-chronological data revealed that there was a phase of deformation in the Lesser Himalaya which took place in pre-Himalayan (Sinha-Roy 1973; Kumar 1997). Seeber and Gornitz (1983), Bhattacharya and Mitra (2009), Mukul (2010) and Matin and Mazumder (2009) suggested that structurally the Proterozoic rock sequence represents a thrust wedge in the foreland propagating thrust system, and these major north dipping thrust systems are associated with tectonic displacement (Yin 2006). The whole Himalayan range comprising major faults/terrace boundary and major thrusts such as Himalayan Frontal Thrust (HFT), Main Boundary Thrust (MBT), Main Central Thrust (MCT), and South Tibetan Detachment (STD), and all these planes have divided Himalaya into several geological units, i.e. Trans Himalayan Granitoids, Tethyan ophiolites, Tethyan Himalayan sequence, Greater Himalayan Sequence, Siwalik rocks (Miocene-Pliocene), Lesser Himalayan Sequence (Proterozoic), and Precambrian basement rocks of Peninsular India (Fig. 2.1).

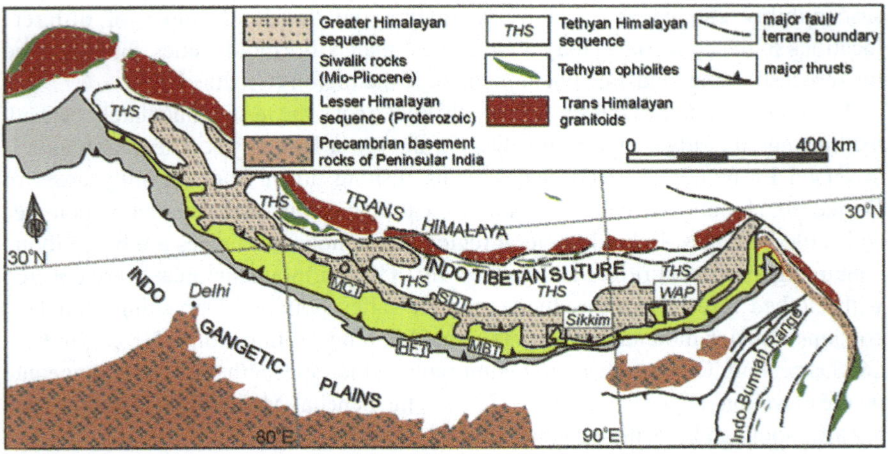

Fig. 2.1 Simplified regional geologic map of the Himalaya (after Acharyya 2007). Study areas in Sikkim and western Arunachal Pradesh (WAP) are shown boxed. *HFT* Himalayan Frontal Thrust, *MBT* Main Boundary Thrust, *MCT* Main Central Thrust, *STD* South Tibetan Detachment

The MCT in Sikkim is marked by a convex to north re-entrant and the MBT is passing through the south of Kurseong having East-West trend (Fig. 2.2). The Daling Group and the Buxa Formation of rocks south of MCT are featured by multiple meso-scale folds and cleavage. There is a distinct variation between the structural pattern in the internal part of the Lesser Himalaya below of MCT, the Higher Himalayan above the MCT, and the Gondwana/Siwalik/Quaternary rocks below the MBT. The Higher Himalayan rock groups are dominated by the younger migmatitic rocks or leucogranites which are existing above the MCT and expressing as the product of the Himalayan Collisional orogeny (Fig. 2.3). The Lesser Himalayan Sequence rocks are associated with older deformation and affected by superimposed deformation.

In Darjeeling-Sikkim Himalaya, heavily compressed lithological composition with microscopic structures is being found below MCT (Fig. 2.4). Such structures help in mechanical and chemical decomposition of rocks rapidly and finally introduce mass movement processes. Basically, Darjeeling-Sikkim Himalaya comprising the lithology of Proterozoic, Upper Palaeozoic-Mesozoic, and Tertiary era (Fig. 2.5). During all these era tectonic activities, i.e. mountain building process took place and developed separate tectono-stratigraphic formation. There are four major lithological groups in Darjeeling Himalaya, i.e. Lingtse gneiss, Gorubathan formation, Reyang formation, and Buxa formation and upper Gondwana.

2.3 Significance of Geomorphic, Geo-tectonic, and Hydrologic Attributes in Landslide Studies

Landslide in the Darjeeling Himalaya is caused as a result of the prevalent of geomorphic, geo-tectonic, and geo-hydrologic factors which helps in downward movement of rock-soil waste under the influence of gravity. All these attributes control

Fig. 2.2 Simplified geological map showing the domal outcrop of the MCT around Rangit window. *MCT* Main Central Thrust, *MBT* Main Boundary Thrust, *MBT*^{U} Roof thrust surrounding Rangit duplex, *GHC* Greater Himalayan Crystallines, *Qal* Quaternary alluvium. Siwalik outcrops are not shown in this simplified map. Asterisks for sample locations (After Roy 1976; Acharyya 2007)

the spatial location, patter, type and distribution of landside phenomena in Darjeeling Himalaya. Geomorphic parameters, i.e. slope, aspect, curvature, relief, geomorphology, soil, land use, and land covers are some contributing factors for landslides occurrences. Steep *slope* is very high susceptible to landslide in comparison to gentle slope. But, the moderate slope with adequate amount of rock-soil debris is more susceptible to slope failure. The occurrences of landslides of a particular location in

Fig. 2.3 Mesoscopic structures in GHC rocks above MCT, Darjeeling-Sikkim. (**a**) Biotite-granite gneiss (politic migmatite) around Gayabari (Darjeeling dt.) showing asymmetric to overturned folds in leucosomal bands, above MCT. (**b**) Tightly folded leucosome with garnet porphyroblasts (brown spots) (After Roy 1976; Acharyya 2007)

Darjeeling Himalaya are also being governed by the existing *slope aspects*. Rainfall is one the most significant triggering factor for causing landslides. Most of the landslide locations are being found in the southern escarpment slope, and it is assumed that south, south-east, and south-west facing slope are most landslide-prone areas which receives huge amount of orographic rainfall and make the slope landslide prone. Very high positive and negative *slope curvature* invites drainage concentration over the space whereas the straight slope did not provide suitable condition for drainage concentration. So, the slope saturation is also governed by the nature of slope curvature and very high positive and negative slope curvature areas are being saturated well which promote down slope movement of slope materials. The areas having maximum differences between higher and lower *elevation* are registered with high landslide susceptibility. Maximum difference in elevation is revealed by maximum slope length, surface run-off, and slope saturation which influences stress properties of the soil and induce slope instability. The areas having high forest density are less susceptible to landslides. Open forest, degraded forest, and bared surfaces are characterized by high landslide susceptibility. The slope with high concentration of human settlements put enormous pressure over the mountain slope and makes the slope susceptible to landslide. To assess landslide susceptibility, land use and land cover are to be studied systematically. Texture, pore space, porosity, permeability, water holding capacity, particle density, and bulk density are regarded as important soil physical properties which control soil saturation, cohesion,

Fig. 2.4 Mesoscopic structures in Darjeeling-Sikkim, below MCT. (**a**) Early inclined to reclined folds in Daling metapsammite below MCT, off Tindharia (After Roy 1976; Acharyya 2007)

internal friction, and shear stress of the slope materials. The existence of various mineralogical compositions in the soil control liquefaction character of the soil and changes cohesion when it comes into contact with water. Chemical properties such as the presence of humus, concentration of sodium, concentration of potassium, and acidity and alkalinity also control the liquefaction properties, i.e. liquid limit, plastic limit, and plasticity index of the soil. The chemical reactions within the soil are being controlled by temperature and rainfall character of the region. Chemical reactions become much faster during monsoon and when slope become much susceptible to landslip.

Geology or lithological composition of the concerned study varies from one place to another. The disintegration and decomposition of the rocks occur in a different way by physical and chemical weathering processes. Different lithological composition responds in a different way when it is exposed to atmospheric processes. The identification of various lithological compositions and preparation of a lithological map were made and it was incorporated to prepare landslide susceptibility map of Darjeeling Himalaya. The study area was dominated by large-scale folding and faulting activities due to active tectonic forces operated in the mountain building processes in the Himalayan region. The existence of thrust plane, bedding plane,

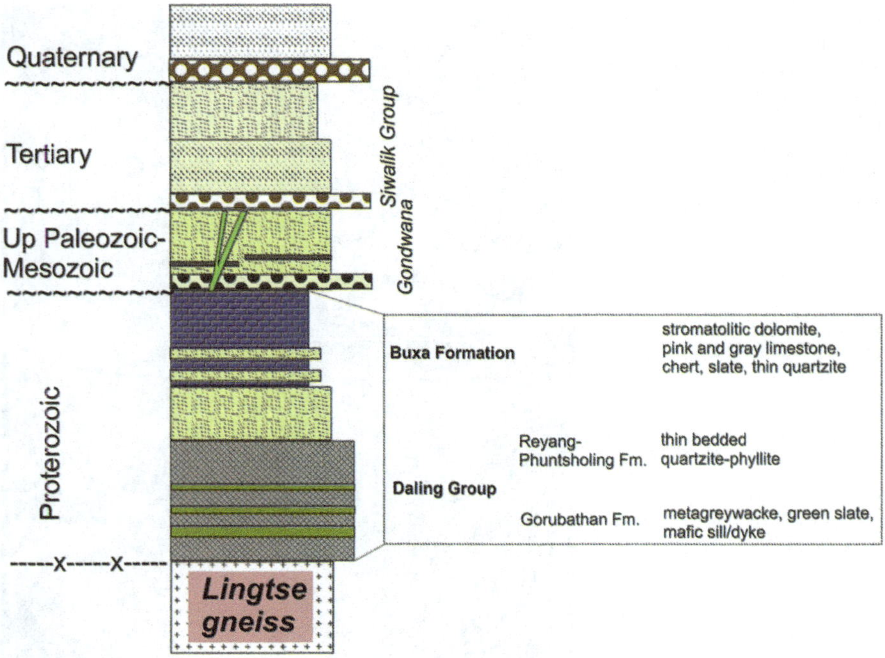

Fig. 2.5 Simplified log showing an overview of tectono-stratigraphic framework, Darjeeling-Sikkim, Eastern Himalaya (After Roy 1976; Acharyya 2007)

and fault are quite common in Darjeeling Himalaya which have played a pivotal role in causing destructive landslide events. It was observed that most of the landslides are located very close to weaknesses plane. The identification of lineaments/weaknesses plane and the preparation of lineament density map were essential for the preparation of landslide susceptibility map. Darjeeling Himalaya is located in a very high to moderate magnitude earthquake region. The vibration and tremor increased shear stresses/driving forces within the disintegrated and decomposed slope materials and finally make the mountain slope more susceptible to landslide. The spatial distribution of earthquake magnitude map is essential for the preparation of landslide susceptibility map.

To prepare landslide susceptibility map, data set regarding distribution of landslides and the causative factors of landslide is very much important (Ermini et al. 2005; Catani et al. 2005; Ercanoglu 2005). Earlier several attempts have already been made to prepare landslide susceptibility zonation map considering various landslide conditioning factors like slope, land cover, precipitation, elevation, distance to road, distance to drainage (Rozos et al. 2011), geology, distance from lineament, slope aspect (Rozos et al. 2011; Mandal and Maiti 2013), slope curvature, drainage density, settlement density (Mandal and Maiti 2013), stream power index, and topographic wetted index (Pourghasemi et al. 2012). In this study, factors like elevation, slope aspect, slope angle, slope curvature, geology, soil, lineament density, distance to lineament, drainage density, distance to drainage, stream power index (SPI), topographic wetted index (TWI), rainfall, normalized differential vegetation index (NDVI) and land use and land cover (LULC) were studied in detail and the landslide probability for each class of landslide inducing factors were also derived. Intensive field investigation was conducted to understand landslide influencing factors prevalent in the concerned study area. Not only that several literatures on landslides activities in Darjeeling Himalayan region were also taken into account to assess landslide conditioning factors and to prepare thematic data layers.

2.4 Generation of Various Thematic Data Layers

2.4.1 Causative Thematic Data Layers Prepared from SRTM DEM

The preparation of thematic data layers which are associated with landside activities were being made scientifically with the help of SRTM DEM and GIS software. Elevation, slope angle, slope aspect, slope curvature, SPI and TWI map of the Darjeeling Himalaya were derived from the SRTM DEM. Both the slope and altitude are positively related to slope instability and have a large impact on the occurrence of landslides in the Himalayas (Singh et al. 2012). Slope and elevation zones were classified into ten. Slope aspect indicates the exposure to atmospheric process, i.e. solar radiation, wind, and rainfall which control the soil moisture availability and promote landslide occurrences (Magliulo et al. 2008). The slope aspect map of the Darjeeling Himalaya was categorized into nine classes, i.e. north, south, east, west, northeast, southeast, northwest, southwest, and flat. The prepared curvature map was classified into three classes, i.e. convex, flat, and concave which indicate slope gradient or slope direction, usually in a particular direction. The SPI is the measure of erosive power of the overland flow based on the assumption that discharge is proportional to the catchment area and predicts the net erosion in the flow acceleration and convergence zones and net deposition in the zones of decreasing flow velocity (Pourghasemi et al. 2012). TWI refers to the accumulation of flow at any point within a watershed. SPI and TWI were derived after Moore et al. (1991).

$$\text{Stream Power Index, SPI} = \text{Ln}\left(A_s^{*} \tan\beta\right) \quad \text{(Eq. 2.1)}$$

$$\text{Topographic Wetness Index, TWI} = \text{Ln}\left(A_s / \tan\beta\right) \quad \text{(Eq. 2.2)}$$

where Ln is the natural log, A_s is the flow accumulation, and $\tan\beta$ is the slope.

2.4.2 Causative Thematic Data Layers Prepared from Landsat TM Imagery

Land use and land cover (LULC), NDVI, lineament number density, distance to lineaments were derived from Landsat TM imagery. The NDVI was estimated using the formula NDVI = (IR − R)/(IR + R), where IR is the infrared portion and R is the Red portion of the electromagnetic spectrum. The positive values of NDVI indicate vegetation health whereas negative values represent water depth.

2.4.3 Causative Thematic Data Layers Prepared from Disaster Mitigation Map

Topographical maps and then updated using Google earth (2015). Drainage density and distance to drainage map was prepared from disaster mitigation map of Darjeeling district using http://darjeeling.gov.in/gismaps.html link. Drainage density (DD) indicates the closeness of spacing between channels (Javed et al. 2009).

$$\mathrm{DD} = L / A \tag{Eq. 2.3}$$

where L = Total length of streams, A = Area of watershed.

2.4.4 Causative Thematic Data Layers of Geology, Soil, and Rainfall

Geology played an important role in slope instability in Darjeeling Himalaya. Geological map was collected from Geological Survey of India (GSI, Kolkata), and the map was grouped into six, i.e. alluvium, Baxa Series, Daling Series, Damuda (Gondwana), Darjeeling Gneiss and Nahan Group (Tertiary). The soil textural map based on soil taxonomy classification was prepared by NBSS (National Bureau of Soil Science, Kolkata) and LUP (Land Use Planning, Kolkata) Regional Centre, Kolkata where mainly nine soil types were identified, i.e. Tista river, W001, W002, W003, W004, W006, W007, W008 and W009 of the Darjeeling Himalaya. Rainfall is one of the most important triggering factors of landslide susceptibility mapping. Rainfall map was prepared based on the data collected from the http://www.worldclim.org website and classified into ten groups accordingly.

2.4.5 *Landslide Inventory Map and Frequency Ratio*

The main basis for landslide susceptibility zonation mapping is to develop a landslide inventory map which can be useful for assessing and reducing landslide hazard or risk on a local scale. Darjeeling Himalaya covering an area of about 2320.313 sq. km with 2,578,126 number of pixels (30 m × 30 m). A total number of 2079 landslides were identified in the study area covering an area of about 6.47 Sq. Km. The largest and smallest landslides cover an area of about 599634 and 1.71 Sq. m respectively. The average area of landslides was 3112.99 Sq. m. with standard deviation of 18721.25 Sq. m. To estimate total landslide affected area and landslide, inventory map of Darjeeling Himalaya Google earth imagery, toposheet, and GPS field survey (2015–2016) were incorporated with full care (Fig. 2.6). In the present study, the frequency ratio for each class of landslide causative factors was derived to assess landslide probability (Eq. 2.4). The frequency ratio showed landslide probability of a specific location, particular class of the causative factors, and the relative significance of each factor in landslide susceptibility assessment and prediction.

$$\mathrm{Fr}_i = \frac{\{N\mathrm{pix}(\mathrm{Si}) / N\mathrm{pix}(\mathrm{Ni})\} * 100}{\{\Sigma N\mathrm{pix}(\mathrm{Si}) / \Sigma N\mathrm{pix}(\mathrm{Ni})\} * 100} \quad \text{(Eq. 2.4)}$$

where N pix (Si) is the number of pixels containing landslide in each class (i), N pix (Ni) is the total number of pixels having class (i) in the whole basin, $\sum N$ pix (Si) total number of pixels in each class, and $\sum N$ pix (Ni) total number of pixels in the whole basin.

2.5 Elevation Character of the Darjeeling Himalaya and Frequency Ratio

The elevation of the Darjeeling Himalaya is classified into ten categories where elevation ranges from 130 to 3600 m. Extreme northwestern part and northeastern part of the Darjeeling Himalaya is dominated by higher elevation which ranges from 1900 to 3500 m. Southern most section, the Tista River and its surrounding area and the Balason river basin are characterized by low elevation and covering large part of the Darjeeling Himalaya. Moderate elevation zones are being found in the eastern and western part of the Darjeeling Himalaya. The maximum area is registered with low elevation zones which is followed by moderate and higher elevation (Table 2.1 and Fig. 2.6). The estimated frequency ratio value for each elevation zones depicted that moderate to low elevation zones having high frequency

Table 2.1 Elevation (in metre) and frequency ratio

Class	Number of pixel (30 × 30 m)	Total pixel (%)	Number of landslide occurrence pixels	Landslide occurrence pixels (%)	Frequency ratio value
130–470	409,786	15.89	465	9.53	0.60
470–770	437,151	16.96	1519	31.11	1.83
770–1100	437,588	16.97	1259	25.78	1.52
1100–1300	246,682	9.57	445	9.11	0.95
1300–1600	342,538	13.29	465	9.53	0.72
1600–1900	260,070	10.09	178	3.65	0.36
1900–2200	171,257	6.64	268	5.50	0.83
2200–2600	142,581	5.53	207	4.25	0.77
2600–3000	80,867	3.14	69	1.42	0.45
3000–3600	49,605	1.92	6	0.12	0.06

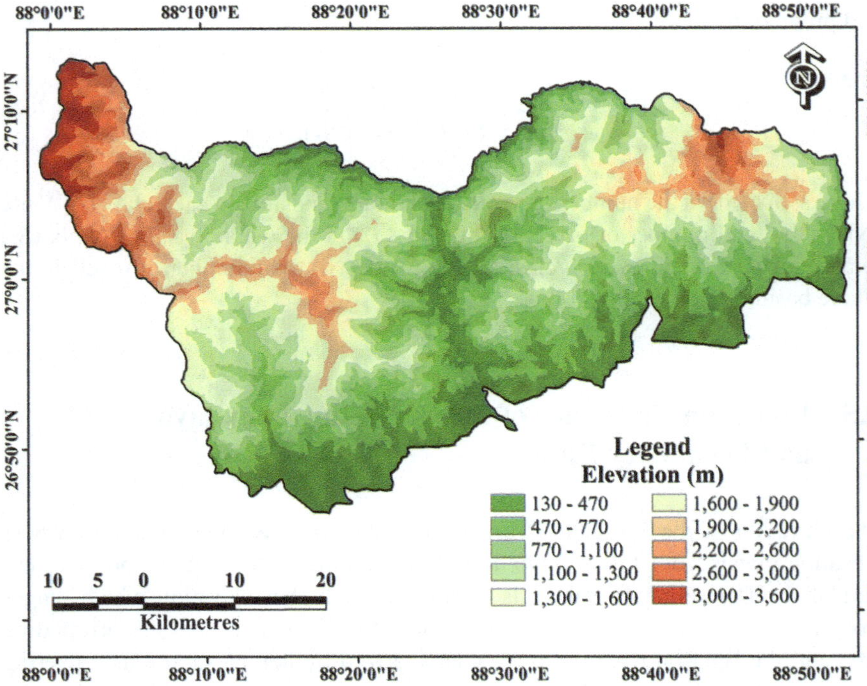

Fig. 2.6 Elevation zonation map of Darjeeling Himalaya

ratio and high probability of landslides. The major valley side slope is characterized by low elevation covered with disintegrated and decomposed materials. As a result of which valley sides slope is characterized by slope movement processes and occurrences of landslide phenomena.

2.6 Slope Angle, Slope Aspect, and Slope Curvature Character of the Darjeeling Himalaya

The slope map of the Darjeeling Himalaya was made from the DEM, and it was classified into ten different slope zones (Fig. 2.7). Basically, the Darjeeling Himalaya is attributed with maximum slope angle where slope angle ranges from 0° to 42.553°. Darjeeling Himalaya is dissected by well-developed drainage network which made the slope steepened by continuous branching of the drainage and its headward erosion. The slope plays an important role for the growth and development drainage network, surface run-off, and soil erosion as well as drainage concentration over the space. Not only that the stream power index (SPI) and topographic wetness index (TWI) are influenced by slope steepness. The study revealed that there is a positive relationship between slope and landslide frequency ratio. The slope having greater than 22° is attributed with the frequency ratio value ranges from 1.09 to 2.47 which showed high probability of landslide phenomena. Moderate slope is also registered with the frequency ratio value of more than 0.9 (Table 2.2).

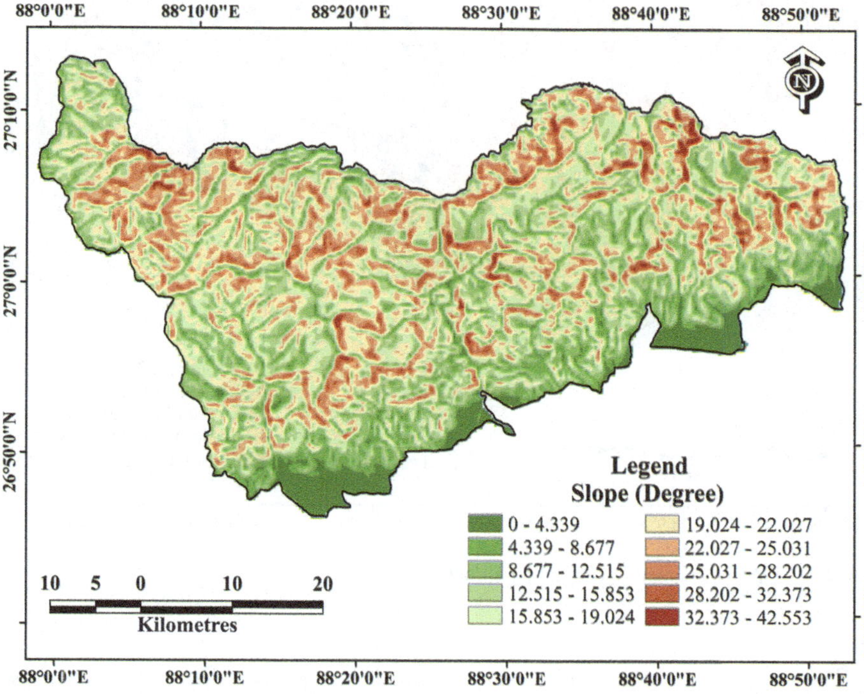

Fig. 2.7 Slope zonation map of Darjeeling Himalaya

Table 2.2 Slope (Degree) character of the Darjeeling Himalaya and landslide probability

Class	Number of pixel (30 × 30 m)	Total pixel (%)	Number of landslide occurrence pixels	Landslide occurrence pixels (%)	Frequency ratio value
0–4.339	187,019	7.25	134	2.74	0.38
4.339–8.677	255,476	9.91	403	8.26	0.83
8.677–12.515	343,632	13.33	590	12.08	0.91
12.515–15.853	374,691	14.53	668	13.69	0.94
15.853–19.024	427,124	16.57	746	15.27	0.92
19.024–22.027	382,230	14.83	704	14.42	0.97
22.027–25.031	278,609	10.81	573	11.73	1.09
25.031–28.202	189,918	7.37	459	9.39	1.28
28.202–32.373	107,393	4.17	501	10.27	2.47
32.373–42.553	32,035	1.24	104	2.14	1.72

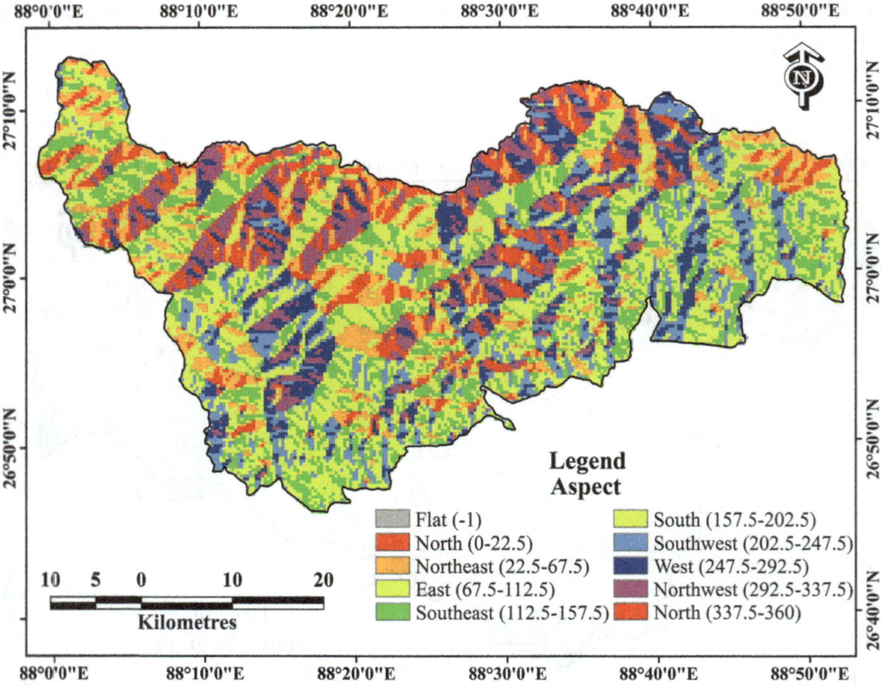

Fig. 2.8 Slope aspect map of Darjeeling Himalaya

The study area is experienced with several slope aspects of flat, north, northeast, east, southeast, south, southwest, west, northwest, and north facing slope (Fig. 2.8). Each slope having individual geomorphic characteristics in terms of soil erosion, run-off, drainage, sediment transport, slope angle, etc. All these geomorphic parameters are being influenced by micro-climatic variation throughout the region. South, southwest, and southeast facing slope received huge amount of orographic rainfall and helps to introduce drainage branching all over the slope and promote surface run-off and soil erosion process. Not only that the disintegration and decomposition of slope are caused all the slope facets due to prevalent of diurnal maximum solar radiation. The estimated frequency ratio value depicted that north-east, east, southeast, south, and southwest facing slope have high probability of landslide occurrences with frequency ratio value of 1.26, 1.19, 1.14, 1.04, and 1.49, respectively. On the other hand, slope aspect of north, west, and northwest are experienced with low frequency ratio and low probability of landslides (Table 2.3).

Slope curvature is another important parameter for landslides in the mountain environment of Darjeeling Himalaya. Three important slope segments are convex slope, straight slope, and concave slope (Fig. 2.9). Amongst all these three, convex and concave slope segments provide a suitable environment for drainage concentration over the space. High positive and high negative curvature values showed high slope convexity and high slope concavity. Such drainage concentration along the convex slope and concave slope make the slope saturate and increase the wetted index value. On the other hand, the saturation leads to less cohesion and internal friction of the slope soil. The places with negative curvature value of –0.095 to –0.644 having the frequency ratio value of more than 1.15. It is also found that the areas with positive curvature value of 0.185 to 0.756 is characterized by high frequency ratio value of more than 0.9 (Table 2.4).

Table 2.3 Slope aspect and frequency ratio (Landslide probability)

Class	Number of pixel (30 × 30 m)	Total pixel (%)	Number of landslide occurrence pixels	Landslide occurrence pixels (%)	Frequency ratio value
Flat (−1)	231	0.01	0	0.00	0.00
North (0–22.5)	136,124	5.28	90	1.84	0.35
Northeast (22.5–67.5)	296,730	11.51	706	14.47	1.26
East (67.5–112.5)	395,973	15.36	890	18.22	1.19
Southeast (112.5–157.5)	413,694	16.05	894	18.30	1.14
South (157.5–202.5)	386,071	14.97	777	15.91	1.06
Southwest (202.5–247.5)	299,310	11.61	844	17.29	1.49
West (247.5–292.5)	266,605	10.34	396	8.11	0.78
Northwest (292.5–337.5)	250,427	9.71	252	5.16	0.53
North (337.5–360)	132,961	5.16	34	0.70	0.14

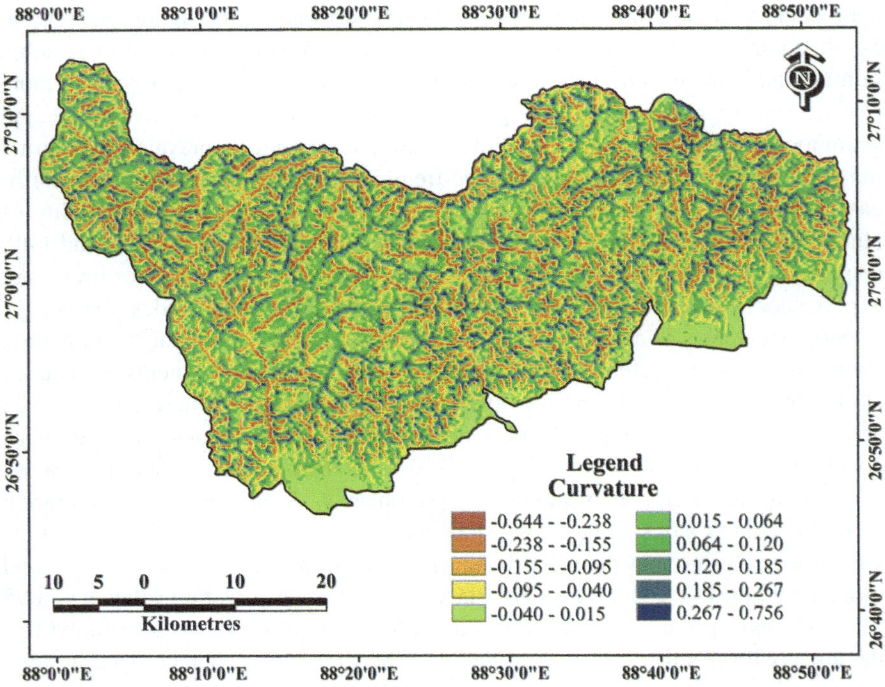

Fig. 2.9 Slope curvature map of Darjeeling Himalaya

2.7 Geology of Darjeeling Himalaya and Landslide Probability

Mallet divided the whole Darjeeling Himalaya into six major categories viz., alluvium, Baxa Series, Daling Series, Damuda (Gondwana), Darjeeling Gneiss and Nahan Group (Tertiary). All these younger formation appears to underlie the older. The gneiss covered the large part of the Darjeeling Himalaya where it varies from a foliated granitoid rock composed to quartz, feldspar, and biotite to a more or less pure mica schist (Fig. 2.10). The daling series consists of phyllite, slate, and quartzite which are being found in the northern and eastern part of the districts. The Buxa series consists of slates, quartzites, and dolomite. The Gondwana system composed mainly of sandstone, shale, and coal and all are characterized by intensively crushed and faulted. As a result of crushing affects, all these rock types were being metamorphosed and converted into quartzites, slates, and graphitic schists. The tertiary beds are composed of soft, massive sandstone containing mica and feldspar.

Darjeeling gneiss covered the maximum area of Darjeeling Himalaya, and this lithological group has high potentiality of landslide occurrences. In Darjeeling Himalaya, Baxa series, Damuda series, and Daling series have high frequency ratio as well as high probability of landslide (Table 2.5).

Table 2.4 Slope curvature and frequency ratio (Landslide probability)

Class	Number of pixel (30 × 30 m)	Total pixel (%)	Number of landslide occurrence pixels	Landslide occurrence pixels (%)	Frequency ratio value
−0.644 to −0.238	48,330	1.87	159	3.26	1.74
−0.238 to −0.155	168,473	6.53	765	15.66	2.40
−0.155 to −0.095	286,017	11.09	855	17.51	1.58
−0.095 to −0.040	390,224	15.14	859	17.59	1.16
−0.040 to 0.015	577,690	22.41	672	13.77	0.61
0.015–0.064	406,708	15.78	548	11.23	0.71
0.064–0.120	330,604	12.82	435	8.91	0.69
0.120–0.185	234,280	9.09	330	6.75	0.74
0.185–0.267	108,702	4.22	192	3.94	0.93
0.267–0.756	27,099	1.05	67	1.38	1.31

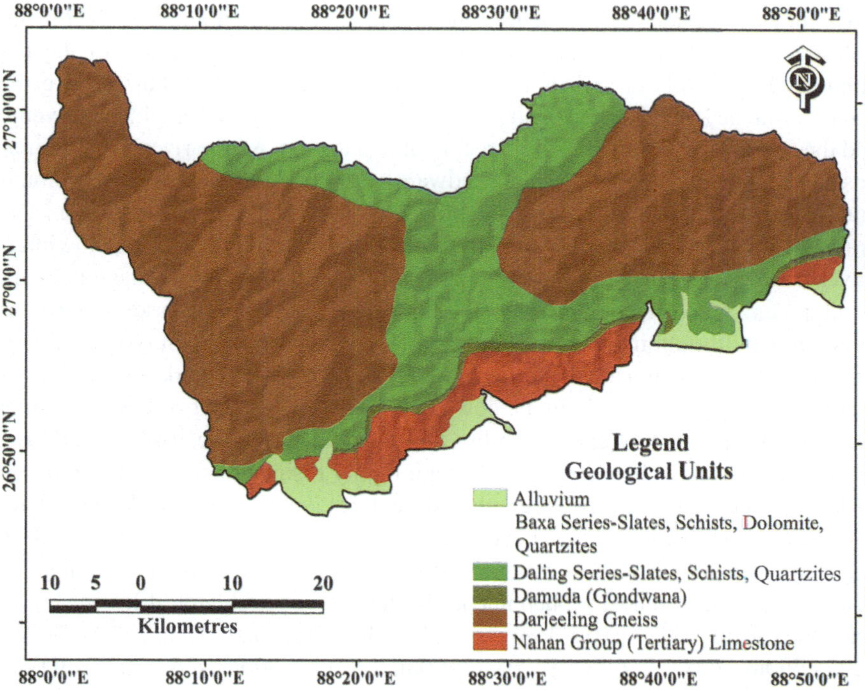

Fig. 2.10 Geological units of Darjeeling Himalaya

Table 2.5 The geology and frequency ratio (Landslide probability) of Darjeeling Himalaya

Name	No. of pixels (30 × 30 m)	Total pixel (%)	No. of landslide occurrence pixels	Landslide occurrence pixels (%)	Frequency ratio value
Alluvium	1,522,044	4.06	27	0.55	0.14
Baxa Series-Slates, Schists, Dolomite, Quartzites	104,522	0.09	4	0.08	0.84
Daling Series-Slates, Schists, Quartzites	2336	27.62	1975	40.46	1.46
Damuda (Gondwana)	36,195	1.40	116	2.38	1.71
Darjeeling Gneiss	713,577	59.09	2631	53.90	0.91
Nahan Group (Tertiary) Limestone	199,453	7.74	128	2.63	0.34

2.8 Distance to Drainage and Drainage Density

The valleys of the southeastern part of the Singalila range are drained by the Mechi, Balason, and Mahanadi River. The remaining part of the Darjeeling Himalaya is dominated by the Tista and its tributaries. The left hand side tributaries are the Rangpo and the Rilli and right hand side tributaries are the Great Rangit, the Rayeng, and the Sivok. The Tista River is a large mountain torrent. All the rivers are continuously engaged in branching through headward erosion as well as changed the mountain slope segment with the help of the processes of erosion and transportation. The drainage network development on both sides of the slope segments make the interfluves area more steep resulting mountain slope instability. The drainage network development, slope steepening, and slope failure are regular processes which promotes an equilibrium in the geomorphic system. The most common types of drainage pattern in the region are dendritic, parallel, etc. The first, second, and third flows rapidly and engaged actively in the down cutting processes. Such valley deepening processes lead to side slope instability and landslides. All the first, second, and third order streams when they pass through highly disintegrated and decomposed rocks, it reduces the cohesive strength and liquefies the slope materials and finally introduces slope movement. The concentration of drainage network and their engagement in the process of erosion and transportation make the mountain slope more vulnerable to landslip phenomena. In the present study, ten different buffer categories were made i.e. 0 – 10m, 10 – 30m, 30 – 70m, 70 – 150m, 150 – 300m, 300 – 700m, 700 – 1200m, 1200 – 1700m, 1700 – 2300m and 2300 – 3000m (Fig. 2.11). Distance up to 10 m from the stream has the greatest probability of landslide phenomena which was characterized by frequency ratio value of 0.175 followed by 10 – 30, 30 – 70, 150 – 300, 70 – 150, 300 – 700, 1200 – 1700, 700 – 1200, 1700 – 2300 and 2300 – 3000 class representing decreasing probability of landslides events respectively. In general, the chance of slope failure events was decreasing from the nearest stream area to farthest stream area (Table 2.6).

The drainage density varies from one place to another based on the developed drainage network in Darjeeling Himalaya. Higher the drainage density more is the

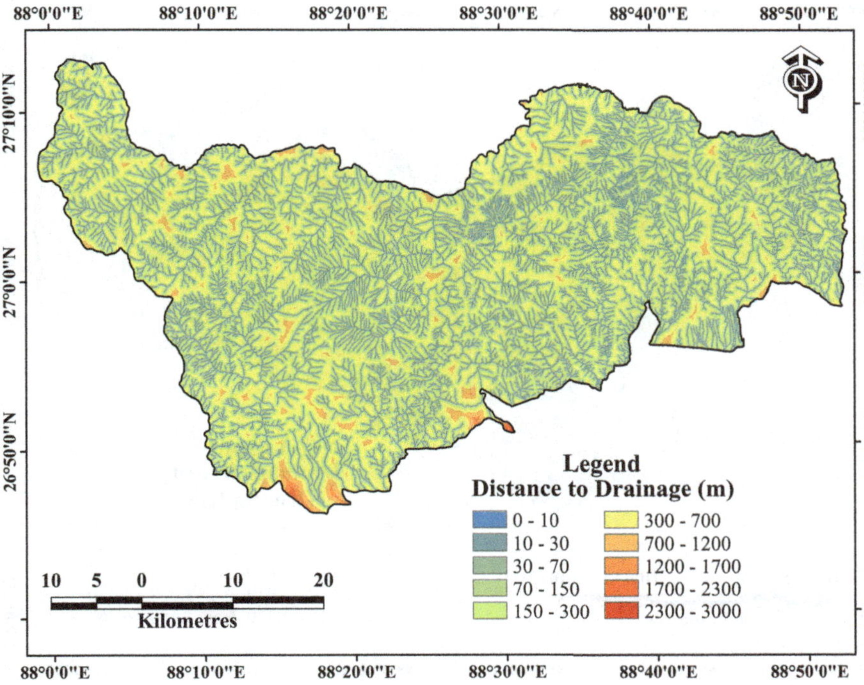

Fig. 2.11 Distance to drainage map of Darjeeling Himalaya

Table 2.6 Distance to drainage and frequency ratio

Class	Number of pixel (30 × 30 m)	Total pixel (%)	Number of landslide occurrence pixels	Landslide occurrence pixels (%)	Frequency ratio value
0–10	92,856	3.60	210	4.31	1.20
10–30	184,399	7.15	412	8.43	1.18
30–70	356,416	13.82	754	15.44	1.12
70–150	628,426	24.38	1291	26.45	1.09
150–300	763,840	29.63	1599	32.76	1.11
300–700	498,808	19.35	600	12.29	0.64
700–1200	45,871	1.78	11	0.23	0.13
1200–1700	5659	0.22	4	0.08	0.38
1700–2300	1593	0.06	0	0.00	0.00
2300–3000	259	0.01	0	0.00	0.00

drainage concentration and slope saturation over the space. It was observed that moderate level of drainage density has high frequency ratio and greater probability of slope instability. High drainage density is being found over the upper part of the mountain slope where branching of drainage and its extension is quite common (Fig. 2.12). But in the middle most section of the slope, second- and third-order streams are prevalent where slope is moderate and branching of drainage is not

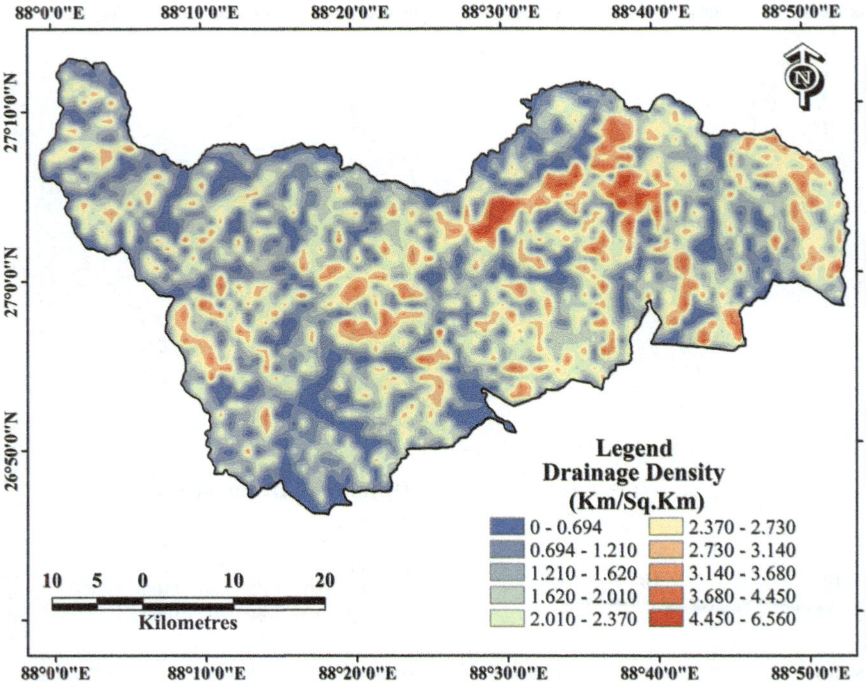

Fig. 2.12 Drainage density map of Darjeeling Himalaya

Table 2.7 The drainage density and landslide probability

Class	No. of pixels	Total pixel (%)	No. of landslide occurrence pixels	Landslide occurrence pixels (%)	Frequency ratio value
0–0.694	243,293	9.44	266	5.44	0.58
0.694–1.210	409,926	15.90	732	14.99	0.94
1.210–1.620	469,031	18.19	879	18.00	0.99
1.620–2.010	498,755	19.35	1202	24.62	1.27
2.010–2.370	393,587	15.27	753	15.42	1.01
2.370–2.730	272,462	10.57	665	13.62	1.29
2.730–3.140	166,766	6.47	279	5.72	0.88
3.140–3.680	76,489	2.97	83	1.69	0.57
3.680–4.450	34,741	1.35	20	0.41	0.31
4.450–6.650	13,076	0.51	4	0.08	0.16

noticeable. As a result, middle most part of the slope is characterized by moderate drainage density but for high slope saturation. In the zone of high drainage density, the frequency ratio ranges from 0.1 to 0.8 (Table 2.7). The low drainage density area is registered with low frequency ratio and low landslide probability. In Darjeeling Himalaya, moderate drainage density area is dominated by maximum landslide occurrences area.

2.9 The Soil of Darjeeling Himalaya and Landslide Probability

The soil of the Darjeeling Himalaya is broadly grouped into two (a) the brown forest soil and (b) the terai soil. The first group of soil is found in the higher elevation zones whereas the second group is found in the lower elevated zone. The brown forest soil is characterized by high fertility and very rich in humus where free drainage is quite common which helps slope saturation and slope instability. The percentage of organic matter gradually decreased down slope. The major types of humus dominated soil found in the forest soils are "Mull" and "Mor". The former is a porous, loose, crumble, and friable mass that develops under deciduous species. The "Mor" is dominated by fungi and mosses and associated to coniferous forest and also display a high degree of saturation of A horizon and little accumulation of sesquioxide in the B horizon (Darjeeling District Gazetteers, 1999). The large part of the Darjeeling Himalaya occupied by coarse loamy humic soil characterized by deep, well-drained, medium run-off, and moderate erosion. Total nine types of soils were seen in the Darjeeling Himalaya i.e. Tista River, W001, W002, W003, W004, W006, W007, W008 and W009 (Table 2.8 and Fig. 2.13).

Different soil types in the Darjeeling Himalaya have the different impacts on future landslide events. The highest probability was found in the W004 soil type category value of 2.33. This soil type was characterized by moderately shallow to deep and well-drained gravelly loamy soil is found in steep slope with moderate permeability, moderately high saturated hydraulic conductivity and moderate erosion covering 40.30% landslide pixels (Table 2.9). This was followed by W001, W002, W008, W003, W007, W006, W009 and Tista river soil type.

2.10 Climate of Darjeeling Himalaya

The Darjeeling Himalaya is attributed with monsoon climate and characterized by warm temperate rainy climate "Cwb" according to Koppen's Climatic Classification. This climate has its own characteristics, i.e. at least ten times as much precipitation in summer month as in the driest winter month such as July rainfall in Darjeeling and Kalimpong are 792.2 and 612.9 mm, respectively, and December rainfall in Darjeeling and Kalimpong are 6.5 and 6.6 mm respectively (Table 2.10). "Cwb" denotes that there is no month with temperature of more than 22 °C whereas at least 4 months having more than 10 °C temperature. The rainfall increases from south to north. The southern Tarai receives an average rainfall amounting 2500 mm and it increases to the north and northeast up to 4500 mm. The maximum average rainfall occurs in the month of June, July, and August where number of rainy days is also high. During these 3 months, the availability of soil moisture prevails for longer duration which helps the liquefaction processes in the soil and makes the slope more vulnerable to landslides.

Table 2.8 Soil characteristics of Darjeeling Himalaya

Symbol	Soil characteristics	Taxonomic name
W001	Shallow, excessively drained, gravelly loamy soils occurring on steep side slopes with gravelly loamy surface, severely eroded associated with rock outcrops	Loamy skeletal, typic Udorthents Rock outcrops
W002	Moderately shallow, excessively drained, coarse loamy soils occurring on steep side slopes with gravelly loamy surface, severe erosion and strong rockiness associated with moderately shallow, well drained, gravelly loamy soils with loamy surface and moderate erosion	Coarse loamy, typic Udorthents Loamy skeletal, typic Dystrochrepts
W003	Deep, well drained, fine loamy soils occurring on steep side slopes with gravelly loamy surface, moderate erosion and moderate rockiness associated with moderately shallow, excessively drained, coarse loamy soils with loamy surface, severe erosion and moderate rockiness	Fine loamy, Umbric Dystrochrepts Coarse loamy, typic Udorthents
W004	Moderately shallow, well drained, gravelly loamy soils occurring on steep side slopes with gravelly loamy surface, moderate erosion and moderate rockiness associated with moderately shallow, somewhat excessively drained, gravelly loamy soil with loamy surface, moderate erosion and moderate rockiness	Loamy skeletal, Typic Haplumbrepts Loamy skeletal, typic Udorthents
W006	Very deep, imperfectly drained, coarse loamy soils occurring on very gently sloping upper piedmont plains with loamy surface associated with very deep, poorly drained, coarse loamy soils	Coarse loamy, Umbric Dystrochrepts Fine loamy, Fluventic Dystrochrepts
W007	Very deep, imperfectly drained, fine loamy soils occurring on very gently sloping lower piedmont plains with loamy surface and moderate erosion associated with very deep, imperfectly drained fine loamy soils	Fine loamy, Fluventic Dystrochrepts Coarse loamy, Aquic Udifluvents
W008	Very deep poorly drained, coarse loamy soils occurring on level to nearly level lower pied, most plain loamy surface associated with very deep, poorly, drained, coarse loamy soils	Coarse loamy, typic haplaquents Coarse loamy, typic fluvaquents
W009	Very deep, imperfectly drained, coarse loamy soils occurring on nearly level lower piedmont plain with loamy surface associated with very deep imperfectly drained, fine loamy soils	Coarse loamy, Aquic Udifluvents Fine loamy, Fluventic Eutrochrepts

2.10.1 Spatial Distribution of Rainfall and Landslide Frequency Ratio

Landslides are closely associated with duration, amount, and intensity of rainfall. The intensity and amount of rainfall varies from one place to another in Darjeeling Himalaya due to topographic configuration. Darjeeling, Kalimpong, Kurseong, Pedong, and Rishi received rainfall differently. The town Kurseong received greater amount of rainfall in comparison to Darjeeling and Kalimpong.

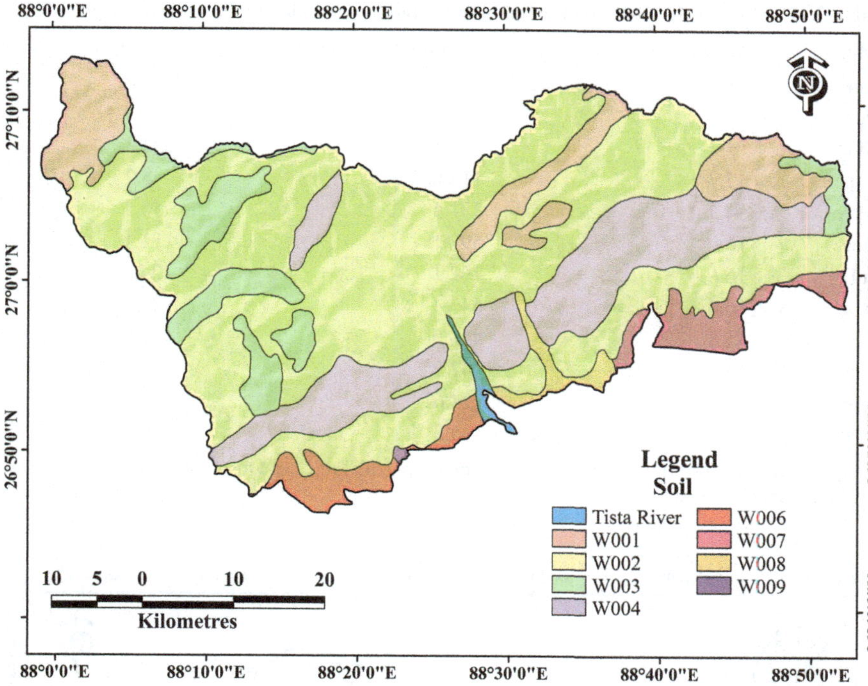

Fig. 2.13 Soil zonation map of Darjeeling Himalaya

Table 2.9 Soil and landslide probability

Class	No. of pixels	Total pixel (%)	No. of landslide occurrence pixels	Landslide occurrence pixels (%)	Frequency ratio value
Tista River	14,861	0.58	0	0.00	0.00
W001	261,470	10.14	492	10.08	0.99
W002	1,434,692	55.65	2183	44.71	0.80
W003	221,855	8.61	183	3.75	0.44
W004	446,612	17.32	1967	40.30	2.33
W006	75,842	2.94	1	0.02	0.01
W007	85,141	3.30	17	0.34	0.10
W008	35,338	1.37	39	0.80	0.58
W009	2314	0.09	0	0.00	0.00

The mean annual rainfall ranges from 1076 to 3705 mm. Based on the rainfall distribution, Darjeeling Himalaya was classified into ten zones (Fig. 2.14). Around 50% landslide affected area and 55% area of Darjeeling Himalaya are characterized by mean annual rainfall of more than 2700 mm. More than 2551 mm rainfall affected area is characterized by high landslide occurrences events, high probabil-

Table 2.10 Monthly distribution of temperature, total rainfall, and number of rainy days

Months	Mean temperature (°C)	Monthly total rainfall (mm)	No. of rainy days
Jan	5.5	10.9	1.0
Feb	5.8	31.7	2.7
Mar	9.7	54.7	4.3
Apr	12.7	113.0	8.2
May	14.3	231.4	15.2
Jun	16.0	597.1	21.1
Jul	16.6	792.1	25.6
Aug	16.6	643.4	23.8
Sep	15.8	445.5	17.0
Oct	13.4	142.2	5.3
Nov	9.7	24.6	1.2
Dec	6.5	6.3	0.7

Source: India Meteorological Department; Climatological Tables of Observatories in India 1960–1991 and Tea Garden data)

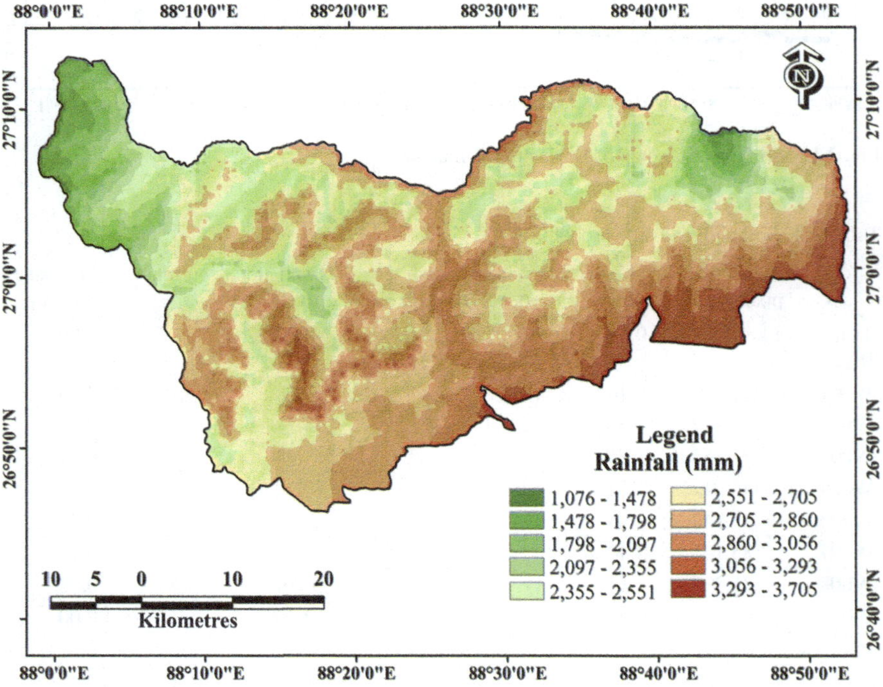

Fig. 2.14 Rainfall Distribution Map of Darjeeling Himalaya

Table 2.11 Mean annual rainfall (mm) and landslide frequency ratio

Class	Number of pixel (30 × 30 m)	Total pixel (%)	Number of landslide occurrence pixels	Landslide occurrence pixels (%)	Frequency ratio value
1076–1478	54,493	2.11	40	0.82	0.39
1478–1798	70,178	2.72	43	0.89	0.33
1798–2097	73,731	2.86	82	1.69	0.59
2097–2355	88,710	3.44	287	5.89	1.71
2355–2551	268,934	10.43	199	4.07	0.39
2551–2705	551,234	21.38	909	18.63	0.87
2705–2860	608,233	23.59	1341	27.48	1.16
2860–3056	423,558	16.43	1165	23.87	1.45
3056–3293	277,977	10.78	628	12.86	1.19
3293–3705	161,077	6.25	187	3.82	0.61

ity, and high frequency ratio (>0.6). The area having mean annual rainfall amount of less than 2000 mm showed low frequency ratio and minimum landslide occurrences phenomena (Table 2.11). The study revealed that the occurrences of landslides and its probability in Darjeeling Himalaya depend upon nature and intensity of rainfall. The distribution of rainfall is associated with slope aspects. South, south-east, and southwest facing slope are dominated by orographic rainfall and all these slope facets become well saturated which increase pore-water pressure and decrease cohesion of the slope materials and finally introduce slope instability. The concave and convex slope segments receive and absorb rainfall more, and these parts encourage slope saturation and down slope movement of disintegrated and decomposed materials. Rainfall in association with slope, aspects, curvature, soil, and lithology plays a significant role in landslide occurrences in Darjeeling Himalaya.

2.11 Topographic Wetness Index (TWI) and Landslide Probability

Topographic wetted index is the expression of slope materials saturation. Higher the value of TI greater is the tendency of slope materials saturation over the space. It is also the expression of slope, upslope contributing area, rainfall of a specific location. Darjeeling Himalaya is classified into ten topographic wetted index zones (Fig. 2.15). The study showed the positive relationship between topographic wetted index and frequency ratio and landslide occurrences area (Table 2.12). More than 85% landslide affected area having high level of topographic wetted index where frequency ratio is also high. The topographic wetted index map of Darjeeling Himalaya

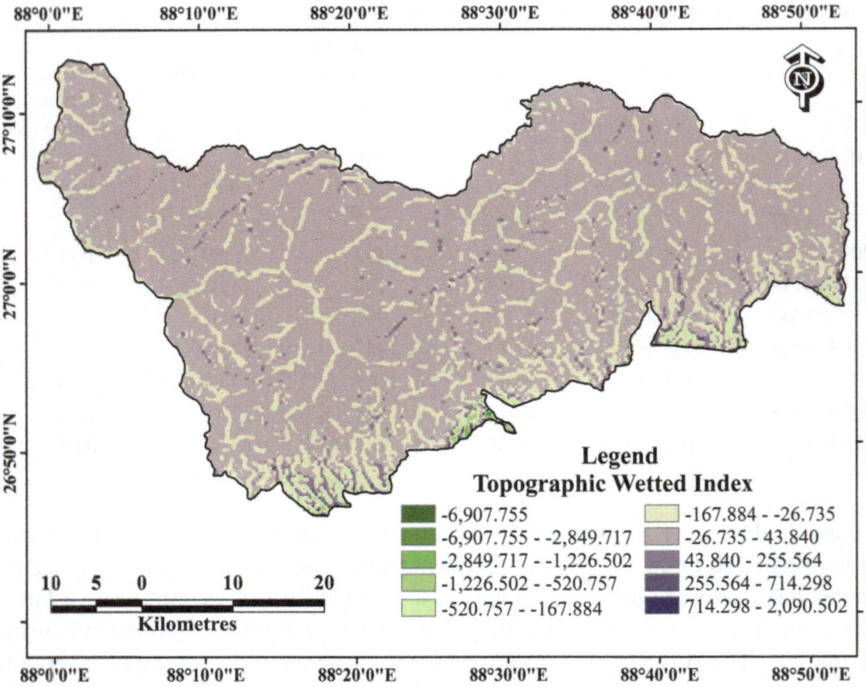

Fig. 2.15 Topographic wetness index map of Darjeeling Himalaya

Table 2.12 Topographic wetted index (TWI) and landslide probability

Class	Number of pixel (30 × 30 m)	Total pixel (%)	Number of landslide occurrence pixels	Landslide occurrence pixels (%)	Frequency ratio value
−6907.755	231	0.01	0	0.00	0.00
−6907.755 to −2849.717	848	0.03	0	0.00	0.00
−2849.717 to −1226.502	2601	0.10	0	0.00	0.00
−1226.502 to −520.757	6953	0.27	0	0.01	0.04
−520.757 to −167.884	36,026	1.40	15	0.32	0.23
−167.884 to −26.735	454,856	17.64	639	13.09	0.74
−26.735 to 43.840	1,987,528	77.09	4150	85.00	1.10
43.840–255.564	76,829	2.98	67	1.37	0.46
255.564–714.298	10,964	0.43	10	0.21	0.49
714.298–2090.502	1289	0.05	0	0.00	0.00

2.12 Stream Power Index (SPI) and Landslide Frequency Ratio

The stream power index analysed the potential flow erosion of a given point of the surface. With increasing of the upslope contributing area, the flow velocity of the streams increases and finally the stream power index, erosion risk as well as movement of slope materials increases. Stream power index values are associated with the factors of conservation of water mass, gravity, basin hydrology, hydraulic geometry, shear stresses, climate, concavity, flood interval, fracture spacing in bedrock, and bedrock erodibility. Slope exhibits an important dominating factor on stream power index and due to which SPI can vary at the reach scale (Fonstad 2003). Stream power index was derived from DEM and it was classified into ten zones (Fig. 2.16). The study revealed that there is a strong positive relationship between stream power index and landslide frequency ratio. The frequency ratio varies from 0.72 to 2.07. Around 20% area of Darjeeling Himalaya is attributed with positive stream power index value where frequency ratio value is more than 1 (Table 2.13). This 20% area is also characterized by 27% landslide affected area out of the total landslide occurrences area. The study envisaged that Darjeeling Himalaya is one of the most significant erosion risk prone area where there is every propensity of slope steepening and down slope movement of weathered materials.

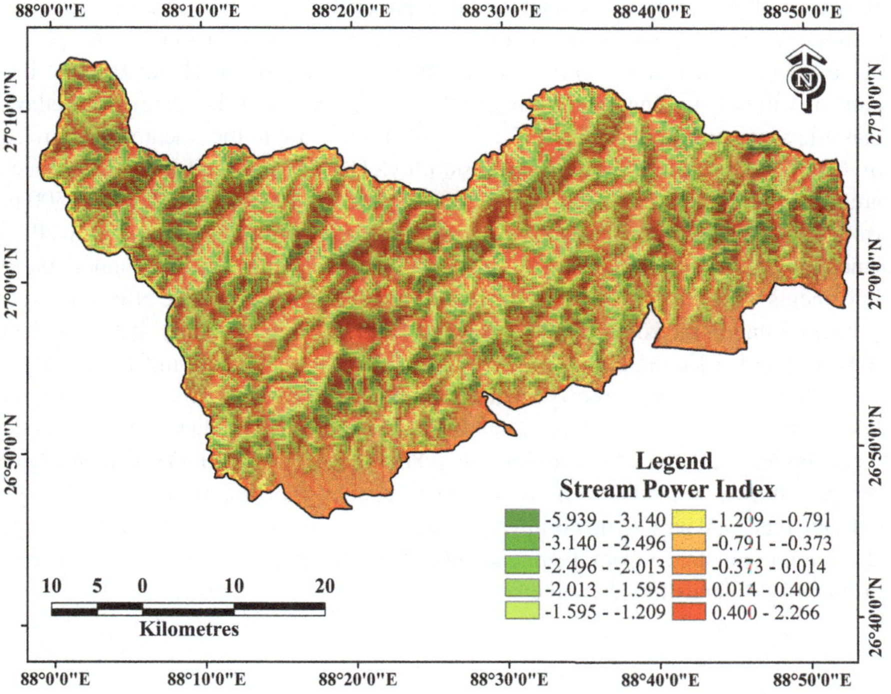

Fig. 2.16 Stream power index of Darjeeling Himalaya

Table 2.13 Stream power index and landslide frequency

Class	No. of pixels	Total pixel (%)	No. of landslide occurrence pixels	Landslide occurrence pixels (%)	Frequency ratio value
−5.939 to −3.140	14,266	0.55	20	0.40	0.72
−3.140 to −2.496	57,082	2.21	92	1.88	0.85
−2.496 to −2.013	132,095	5.12	221	4.53	0.88
−2.013 to −1.595	223,338	8.66	379	7.76	0.90
−1.595 to −1.209	296,386	11.50	482	9.87	0.86
−1.209 to −0.791	361,250	14.01	612	12.55	0.90
−0.791 to −0.373	379,786	14.73	691	14.16	0.96
−0.373 to 0.014	571,397	22.16	1040	21.29	0.96
0.014–0.400	391,010	15.17	752	15.41	1.02
0.400–2.266	151,516	5.88	593	12.15	2.07

2.13 Distance to Lineament and Landslide Frequency Ratio

The lineaments are the weaknesses plane, i.e. fault, crack, joint, bedding plane, thrust plane over the space. The lineaments distribution map of Darjeeling Himalaya showed well-distributed lineaments and depicted the region as one of the active tectonic activities prone segment of the Himalayan mountain range. The places close to lineaments show more slope instability and vice versa. The distance to lineament varies from one location to another and accordingly Darjeeling Himalaya classified into nine zones (Fig. 2.17). The locations close to the lineaments showed high frequency ratio and landslide possibility. The derived landslide occurrences area of more than 70% located within 2000 m from the lineaments. Beyond 3500 m from the lineaments the landslide occurrences area and frequency ratio as well as landslide probability is very less (Table 2.14). The frequency ratio in connection to lineaments varies from 0 to 1.25. The maximum number of landslide affected pixels is being found in the zones of 400–1000 and 1000–2000 m. Not only that more than 60% area of Darjeeling Himalaya existing between 400 and 2000 m from the location of lineaments. It was observed that the locations between 2000 m and 3500 m from lineaments are more vulnerable to landslide phenomena. The distance from lineaments between 20 and 100 m covers only 3.86% area of Darjeeling Himalaya where the probability of landslide is high with frequency ratio of more than 0.8. The extreme marginal part in the northwest, south, and southeast of Darjeeling Himalaya showed more distance from the lineaments where the frequency ratio is low and low probability of landslide phenomena.

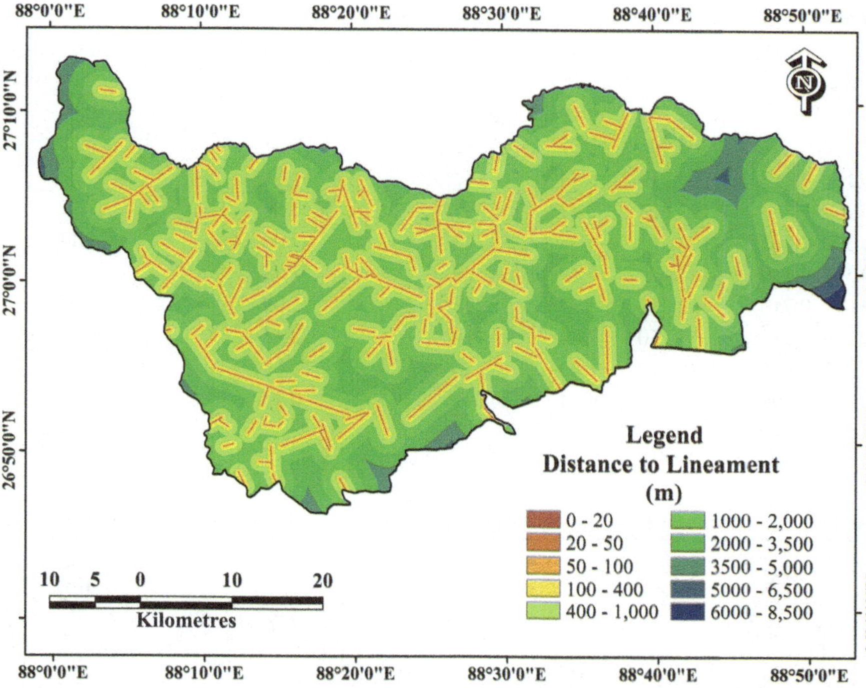

Fig. 2.17 Distance to lineaments map of Darjeeling Himalaya

Table 2.14 Distance to lineaments and landslide frequency ratio

Class	No. of pixels	Total pixel (%)	No. of landslide occurrence pixels	Landslide occurrence pixels (%)	Frequency ratio value
20–50	36,643	1.42	61	1.25	0.88
50–100	63,438	2.46	119	2.43	0.99
100–400	406,582	15.77	823	16.86	1.07
400–1000	787,207	30.53	1426	29.22	0.96
1000–2000	781,676	30.32	1448	29.66	0.98
2000–3500	369,169	14.32	872	17.87	1.25
3500–5000	95,146	3.69	91	1.87	0.51
5000–6500	10,670	0.41	0	0.00	0.00
6500–8500	3904	0.15	0	0.00	0.00

2.14 Lineament Density and Landslide Frequency Ratio

The lineament density in Darjeeling Himalaya ranges from 0.056 to 1.590. Moderate lineament density showed high frequency ratio of 1.11, 1.45, 1.28, and 0.85 as well as high landslide probability (Table 2.15). More than 15% area of Darjeeling Himalaya is characterized by moderate to high density of lineaments. Large part of the Darjeeling Himalaya is covered with low lineament density where the value ranges from 0 to 0.287 (Fig. 2.18).

Table 2.15 Lineament density and landslide frequency ratio

Class	No. of pixels	Total pixel (%)	No. of landslide occurrence pixels	Landslide occurrence pixels (%)	Frequency ratio value
0–0.056	1,309,367	50.79	2325	47.63	0.94
0. 056–0.168	300,996	11.67	482	9.87	0.85
0.168–0.287	230,519	8.94	577	11.81	1.32
0.287–0.399	192,603	7.47	337	6.90	0.92
0.399–0.505	162,361	6.30	342	7.00	1.11
0.505–0.610	141,605	5.49	390	7.98	1.45
0.610–0.716	102,252	3.97	248	5.09	1.28
0.716–0.835	73,150	2.84	118	2.42	0.85
0.835–0.966	48,815	1.89	53	1.08	0.57
0.966–1.590	16,458	0.64	11	0.22	0.35

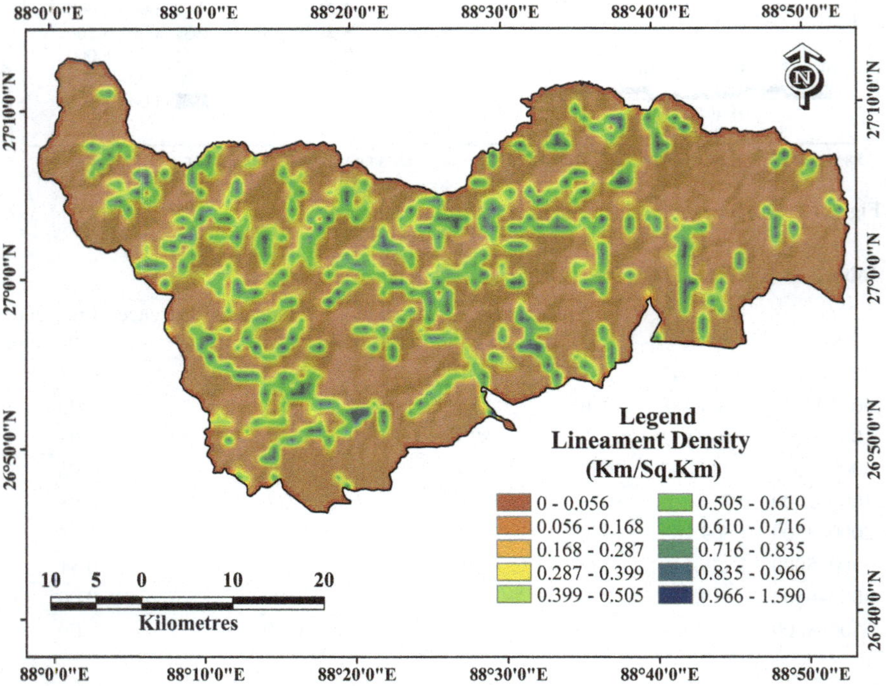

Fig. 2.18 Lineament density map of Darjeeling Himalaya

2.15 Land Use and Land Cover (LULC), Normalized Differential Vegetation Index (NDVI) and Landslide Probability (Frequency Ratio)

Land use and land cover play a significant role in changing the slope material by introducing physical and anthropogenic processes. Destruction forest area leads to the exposure of slope surface to atmospheric processes and weathering processes which helps in movement processes. The growth of plants promotes root-wedging process and helps in disintegration of rocks. The chemical decomposition of dead plants root changes mineralogical composition of slope materials and helps in down slope movement. The expansion of human settlement and agricultural land put enormous pressure on land and slope soil. The Darjeeling Himalaya is classified into nine major land use and land cover, i.e. social forestry, agricultural single crop, forest, rural settlement, urban settlement, agricultural plantation, wasteland dry/barren land, public utility and facility and wastelands with scrub (Fig. 2.19). All these land use and land cover type were incorporated with the landslide inventory of Darjeeling Himalaya and accordingly frequency ratio was estimated. The derived frequency ratio value depicted that landslide probability is high in wasteland dry/barren land

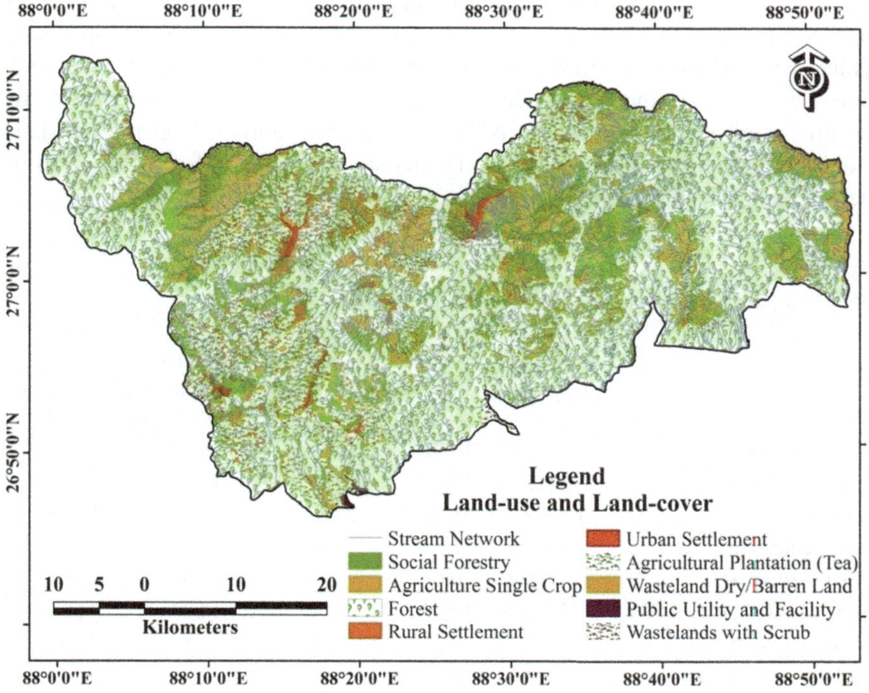

Fig. 2.19 Land use and land cover of Darjeeling Himalaya

(FR 7.50), wastelands with scrub (FR 5.49), forest (FR 1.10), agricultural single crop (FR 0.85) and agricultural plantation (FR 0.80) respectively.The area covered by public utility and facility (FR 0.00), rural settlement (FR 0.63) and urban settlement (FR 0.58) revealed low frequency ratio and low landslide probability respectively (Table 2.16).

The clearance of forest cover and development of terraces along steep introduce a platform for settling run-off from the upslope contributing area. The creation of terraces retards the fast flow of run-off and encourages percolation of water downward. The situation leads to slope materials saturation at greater depth and increases pore-water pressure and reduces cohesion and finally insists slope failure. As a result of which most of the plantation area of Darjeeling Himalaya is dominated by destructive landslide phenomena (Fig. 2.20). NDVI is associated with vegetation density over the space. High vegetation density shows high NDVI and vice versa. In Darjeeling Himalaya, NDVI ranges from −0.220 to 0.667 (Table 2.17). Low NDVI showed minimum vegetation cover and low vegetation density and maximum exposure of the surface to atmospheric process. In Darjeeling Himalaya, the places having low NDVI is associated with high frequency ratio and high probability of landslide. High frequency ratio and high landslide probability are being found at locations where NDVI ranges from -0.220 to 0.263.

All the geo-tectonic, geomorphic, and geo-hydrologic parameters play a significant role in causing landslides in the mountain environment. Not only the factors but also the individual category of all the landslide occurrence factors showed a distinct level of landslide probability in Darjeeling Himalaya. To assess the landslide phenomena, all the factors such as slope angle, slope aspect, slope curvature, elevation, soil, geology, drainage, NDVI, lineaments, rainfall, land use and land cover, topographic wetness index (TWI), and stream power index (SPI) are to be analysed in a scientific way.

Table 2.16 Land use and land cover and landslide probability

Class	Number of pixel	Total pixel (%)	Number of landslide occurrence pixels	Landslide occurrence pixels (%)	Frequency ratio value
Social forestry	309,800	12.02	453	9.27	0.77
Agricultural single crop	331,856	12.87	537	10.99	0.85
Forest	1,430,826	55.50	2978	61.00	1.10
Rural settlement	126,193	4.89	150	3.07	0.63
Urban settlement	16,672	0.62	18	0.36	0.58
Agricultural plantation (Tea)	337,588	13.09	514	10.53	0.80
Wasteland dry/barren land	399	0.02	7	0.15	7.50
Public utility and facility	3052	0.12	0	0.00	0.00
Wastelands with scrub	21,740	0.84	225	4.61	5.49

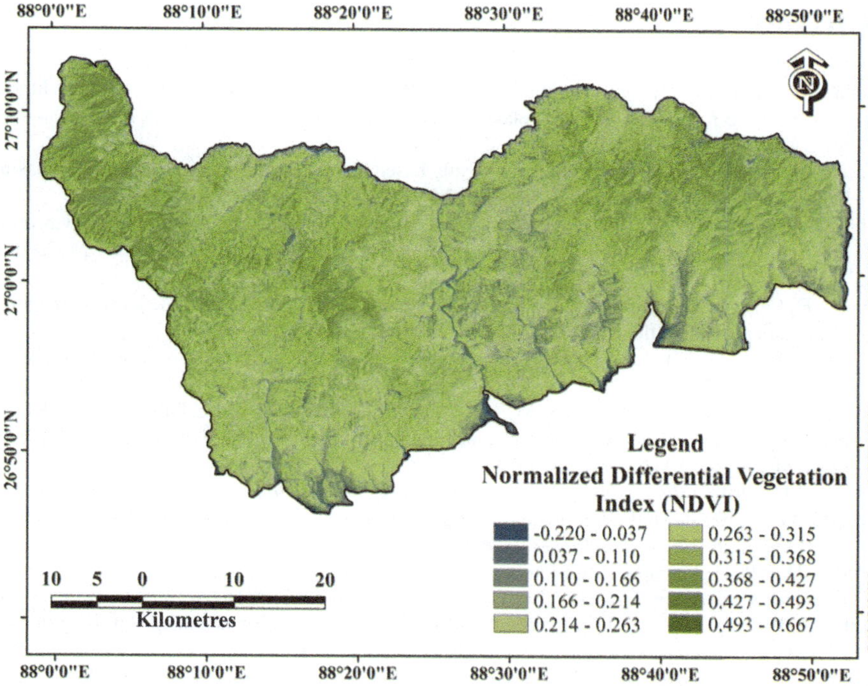

Fig. 2.20 NDVI map of Darjeeling Himalaya

Table 2.17 NDVI and landslide probability

Class	Number of pixel	Total pixel (%)	Number of landslide occurrence pixels	Landslide occurrence pixels (%)	Frequency ratio value
−0.220 to 0.037	34,132	1.32	324	6.64	5.03
0.037–0.110	77,506	3.01	482	9.87	3.28
0.110–0.166	167,231	6.49	520	10.65	1.64
0.166–0.214	273,220	10.60	683	13.98	1.32
0.214–0.263	381,618	14.80	759	15.54	1.05
0.263–0.315	451,331	17.51	753	15.43	0.88
0.315–0.368	428,060	16.60	587	12.02	0.72
0.368–0.427	364,538	14.14	425	8.70	0.62
0.427–0.493	257,616	9.99	245	5.02	0.50
0.493–0.667	142,875	5.54	105	2.15	0.39

References

Acharyya, S. K. (2007). Evolution of the Himalayan Paleogene foreland basin, influence of its litho-packet on the formation of thrust-related domes and windows in the Eastern Himalayas—A review. *Journal of Asian Earth Sciences, 31*, 1–17.

Beven, K. J., & Kirkby, M. J. (1979). A physically based variable contributing area model of basin hydrology. *Hydrological Sciences Bulletin, 24*(1), 43–69.

Bhattacharya, K., & Mitra, G. (2009). A new kinematic evolutionary model for the growth of a duplex—An example from the Rangit duplex, Sikkim Himalaya, India. *Gondwana Research, 16*, 697–715.

Catani, F., Casagli, N., Evmini, L., Righini, G., & Menduni, G. (2005). Landslide hazard mapping at catchment scale in the Arno River basin. *Landslides, 2*, 329–342.

Chandler, R. J. (1972). Lias Clay: Weathering processes and their effect on shear strength. *Geotechnique, 15*, 403–431.

Colombo, A., Lanteri, L., Ramasco, M., & Troisi, C. (2005). Systematic GIS based landslide inventory as the first step for effective landslide hazard management. *Landslide, 2*, 291–301. https://doi.org/10.1007/s10346-005-0025-9.

Crozier, M. J. (1989). *Landslides: Causes, consequences and environment*. London: Routledge.

Crozier, M. J. (1999). Prediction of rainfall triggered landslides: A test of the antecedent water status model. *Earth Surface Processes and Landforms, 24*, 825–833.

Cruden, D. (1991). A simple definition of a landslide. *Bulletin of the International Association of Engineering Geology, 43*, 27–29. https://doi.org/10.1007/BF02590167.

Dehn, M., Burger, G., Buma, J., & Gasparetto, P. (2000). Impact of climate change on slope stability using expanded downscaling. *Engineering Geology, 55*, 193–204.

Dietrich, W. E., Reiss, R., Hsu, M. & Montgomery, D. R. (1995). A process-based model for colluvial soil depth and shallow landsliding using digital elevation data. *Hydrological Processes 9*, 383–400.

Dietrich, W. E., Bellugi, D. & Real de Asua, R. (2001). Validation of the shallow landslide model SHALSTAB for forest management. In M. S. Wigmosta & S. J. Burges (Eds.), *Land Use and Watersheds: Human influence on hydrology and geomorphology in urban and forest areas. American Geophysical Union Water Science and Application* 2, 195–227.

Ermini, L., Catani, F., & Casagli, N. (2005). Artificial neural network applied to landslide susceptibility assessment. *Geomorphology, 66*, 327–347.

Fonstad, M. A. (2003). Spatial variation in the Power of Mountain stream in the Sangre de Cristo Mountains, New Mexico. *Geomorphology, 55*(1–4), 75–96.

Galli, M., Ardizzone, F., Cardinali, M., Guzzetti, F., & Reichenbach, P. (2008). Comparing landslide inventory map. *Geophysical Journal of the Royal Astronomical Society, 94*, 268–269.

Glade, T. (1998). Establishing the frequency and magnitude of landslide-triggering rainstorm events in New Zealand. *Environmental Geology, 35*, 160–174.

Guzzetti, F., Reichenbach, P., Cardinali, M., Galli, M., & Ardizzone, F. (2005). Landslide hazard assessment at the basin scale. *Geophysical Journal of the Royal Astronomical Society, 72*, 272–299.

Javed, A., Ahmad, R., & Khanday, M. Y. (2009). Prioritization of sub-watersheds based on morphometric and land use analysis using remote sensing and GIS techniques. *Jounal of Idian*, 261–274.

Kumar, G. (1997). *Geology of Arunachal Pradesh*. Bangalore: Geological Society of India.

Le Fort, P. (1975). Himalayas the collided range; present knowledge of the continental arc. *American Journal of Science, 275A*, 1–44.

Lee, S. & Sambath, T. (2006). Landslide susceptibility mapping in the Damrei Romel area, Cambodia using frequency ratio and logistic regression models. Environmental Geology, 50, 847–856.

Lee, S., & Pradhan, B. (2007). Landslide hazard mapping at Selangor, Malaysia using frequency ratio and logistic regression models. *Landslides, 4*(1), 33–41.

Mandal, S., & Maiti, R. (2013). Integrating the analytical hierarchy process (AHP) and the frequency ratio (FR) model in landslide susceptibility mapping of Shivkhola Watershed, Darjeeling Himalaya. *International Journal of Disaster Risk Science, 4*(4), 200–212.

Matin, A., & Mazumder, S. (2009). Deformation mechanisms in the frontal Lesser Himalayan Duplex in Sikkim Himalaya, India. *Journal of Earth System Science, 118*, 379–390.

Montgomery, D. R., & Dietrich, W. E. (1994). A physically based model for the topographic control on shallow land sliding. *Water Resources Research, 30*(4), 1153–1171.

Moore, I. D., Grayson, R. B., & Ladson, A. R. (1991). Digital terrain modelling: A review of hydrological, geomorphological and biological applications. *Hydrological Processes, 5*, 3–30.

Mukul, M. (2010). First-order kinematics of wedge-scale active Himalayan deformation: Insights from DarjilingeSikkimeTibet (DaSiT) wedge. *Journal of Asian Earth Sciences, 39*, 645–657.

Magliulo, P., Di Lisio, A., Russo, F., & Zelano, A. (2008). Geomorphology and landslide susceptibility assessment using GIS and bivariate statistics: a case study in southern Italy. *Natural Hazards, 47*(3), 411–435.

Pourghasemi, H. R., Pradhan, B., & Gokceoglu, C. (2012). Application of fuzzy logic and analytical hierarchy process (AHP) to landslide susceptibility mapping at Haraz watershed, Iran. *Natural Hazards, 63*, 965–996.

Preston, N. J. (2000). Feedback effects of rainfall triggered shallow landsliding. In E. Bromhead, N. Dixon, & M. L. Ibsen (Eds.), *Landslide in research, theory and practice* (pp. 1239–1244). Cardiff: Thomas Telford.

Roy, K. K. (1976). Some problems of stratigraphy and tectonics of the Darjeeling and Sikkim Himalayas. *Geological Survey of India, Miscellaneous Publication, 24*, 379–394.

Rozos, D., Bathrellos, G. D., & Skilodimou, H. D. (2011). Comparison of the implementation of rock engineering system (RES) and analytical hierarchy process (AHP) methods, based on landslide susceptibility maps, compiled in GIS environment: A case study from the Eastern Achai County of Peloponnesus, Greece. *Environment and Earth Science, 63*(1), 49–63.

Rybar, J. (1997). Increasing impact of anthropogenic activities upon natural slope stability. In P. G. Marinos, G. C. Koukis, G. C. Tsiambaos, & G. C. Stournaras (Eds.), *Proceedings of the International Symposium on Engineering Geology and Environment, 23-27 June 1997, Athens, Greece* (pp. 1015–1020). Rotterdam: A.A. Balkema.

Schmidt, K. M., Roering, J. J., Stock, J. D., Dietrich, W. E., Montgomery, D. R., & Schaub, T. (2001). The variability of root cohesion as an influence on shallow landslide susceptibility in the Oregon Coast Range. *Canadian Geotechnical Journal, 38*, 995–1024.

Seeber, L., & Gornitz, V. (1983). River profiles along Himalayan arc as indicators of active tectonics. *Tectonophysics, 92*, 335–367.

Shroder, J. F., & Bishop, M. P. (1998). Mass movement in the Himalaya: New insights and research directionns. *Geomorphology, 26*, 13–35.

Singh, Y., Bhat, G. M., Sharma, V., Pandita, S. K., & Thakur, K. K. (2012). Reservoir induced landslide at Assar, Jammu and Kashmir: A case study. *Journal of the Geological Society of India 80*(3), 435–439.

Sinha-Roy, S. (1973). Tectonic evolution of the Darjeeling Himalayas. *Quarterly Journal Geological, Mining and Metallurgical Society (India), 48*, 167–178.

Yin, A. (2006). *Earth-Science Reviews, 76*, 1–13.

Yin, A., Dubey, C. S., Webb, A. A. G., Kelty, T. K., Grove, M., Gehrels, G. E., et al. (2010). Geologic correlation of the Himalayan orogen: part 1. Structural geology, U-Pb zircon geochronology, and tectonic evolution of the Shillong Plateau and its neighbouring regions in NE India. *Geological Society of America Bulletin, 122*, 336–359.

Chapter 3
Frequency Ratio (FR) Model and Modified Information Value (MIV) Model in Landslide Susceptibility Assessment and Prediction

3.1 Introduction

Researchers realized that the proper understanding of landslide phenomena and development of landslide zoning is one of the most significant principles towards landslide management and mitigation strategies. Since then they started to introduce various physical and statistical models in connection to landslide studies. Geoinformatics-based landslide hazard susceptibility assessment and prediction have been studied by Muthu and Petrou (2007); Caiyan and Jianping (2009); Rowbotham and Dudycha (1998); Donati and Turrini (2002); Lee and Pradhan (2006a, b, 2007); Sarkar and Kanungo (2004); Pandey et al. (2008); and Nithya and Prasanna (2010). Guzzetti et al. (1999) suggested that landslide susceptibility mapping can be well studied using multi-variate or bi-variate statistical approach. This approach dealt with the relationship between landslide conditioning factors and location of landslides in a geographical region. Atkinson and Massari (1998) introduced logistic regression model for landslide susceptibility zonation mapping. Jibson et al. (2000) and Zhou et al. (2002) incorporated probabilistic models in assessing landslide hazard and risk. Fuzzy logic and artificial neural network models were used in various studies, i.e. Ercanoglu and Gokceoglu (2004); Pistocchi et al. (2002); Lee et al. (2003, 2004); Pradhan and Lee (2010). Frequency ratio model, artificial neural network model, decision tree model, weights of evidence model, and fuzzy logic were introduced by Bagherzadeh and Mansouri Daneshvar (2012); Ilanloo (2011); Poudyal et al. (2010). Lee and Sambath (2006), Lee and Pradhan (2007), Mandal and Maiti (2011, 2013), and Mandal and Mandal (2016) used frequency ratio model to assess a correlation between the historical slide locations and various influencing factors under consideration. Intarawichian and Dasananda (2011) applied frequency ratio model to analyse slope instability and ascribed the model as a popular quantitative method.

Modefied information value (MIV) model is a bi-variate statistical method for spatial assessment and prediction of landslides which showed the relationships between

S. Mandal, S. Mondal, *Statistical Approaches for Landslide Susceptibility Assessment and Prediction*, https://doi.org/10.1007/978-3-319-93897-4_3

landslide phenomena and landslide influencing factors. The information values for each class of the landslide influencing factors were determined by studying the landside inventory of the Darjeeling Himalaya. Sarkar et al. (2006) presented a GIS-based spatial data analysis for susceptibility mapping of Sikkim Himalayas. They derived numerical weights for each data layers to perform Information Value Model. Zezere (2002) suggested that anthropogenic activities play significant role in landslide occurrence and magnitude of landslides which depends largely upon typology of landslides. A landslide susceptibility maps for Minamata area of Japan developed by using logistic regression and information value model on GIS environment after Wang and Sassa (2005). Sharma et al. (2009) revealed that the accuracy assessment of landslide susceptibility map confirmed the model with highest degree of accuracy for high susceptibility class. Akbar and Ha (2011) opined that information value model is an integrated model for landslide susceptibility zonation mapping using a modified form of pixel-based information value model. In the present study, frequency ratio (FR) model and modified information value (MIV) model were used to assess and predict the spatial distribution of landslide susceptibility in Darjeeling Himalaya as well as to compare between FR and MIV models. Finally, a validation process was adopted to find out the accuracy level for both statistical models (FR and MIV).

3.2 Materials and Methods

To perform frequency ratio model and information value model various data layers, i.e. elevation, slope aspect, slope angle, slope curvature, geology, soil, lineament density, distance to lineament, drainage density, distance to drainage, stream power index (SPI), topographic wetted index (TWI), rainfall, normalized differential vegetation index (NDVI) and land use and land cover (LULC) were prepared using RS and GIS and others secondary data from various sources (Chap. 1, Table 1.7). To derive frequency ratio for each class, all the data layers were incorporated with the prepared landslide inventory map (Fig. 3.3).

3.2.1 Frequency Ratio (FR) Model and Landslide Susceptibility

Frequency ratio model dealt with the number of pixels affected by landslides and number of pixels prevailing in a specific geographical unit of the study. Here, in the landslide susceptibility zonation mapping process each class of landslide conditioning factors and associated pixels with and without landslides were taken into account. This principle was adopted by Lee and Talib (2005), Porghasemi (2007), Lee and Pradhan (2007), Pradhan and Lee (2009, 2010), Avinash and Ashamanjari (2010), Karim et al. (2011), and Intarawichian and Dasananda (2011) to perform "FR" model and to generate landslide susceptibility zonation map of Darjeeling Himalaya. To obtain *frequency ratio (FR)* for each class of all the data layers, a combination has been established between landslide inventory map and criterion maps using the following principle (Karim et al. 2011). To perform frequency ratio model, 15 landslide conditioning factors were taken into account and according frequency ratio value of each class was derived (Fig. 3.1 and Table 3.1).

$$\mathrm{Fr_i} = \frac{N_{\mathrm{pix}(s_i)} / N_{\mathrm{pix}(N_i)}}{\sum N_{\mathrm{pix}(s_i)} / \sum N_{\mathrm{pix}(N_i)}} \qquad \text{(Eq. 3.1)}$$

$N_{\mathrm{pix}(s_i)}$: The number of pixels containing slide in each class (i),
$N_{\mathrm{pix}(N_i)}$: Total number of pixels having class (i) in the whole watershed
$\sum N_{\mathrm{pix}(s_i)}$: Total number of pixels containing landslide
$\sum N_{\mathrm{pix}(N_i)}$: Total number of pixels in the whole area of the watershed

To obtain landslide susceptibility index (LSI), frequency ration of each range/class of all the landslide triggering factors were summed (eq. 3.2) after Pradhan (2010).

$$\mathrm{LSI} = \mathrm{Fr}_1 + \mathrm{Fr}_2 + \cdots + \mathrm{Fr}_n \qquad \text{(Eq. 3.2)}$$

where LSI= Landslide susceptibility index; Fr= Frequency ratio/rating to each class/range of landslide triggering factor.

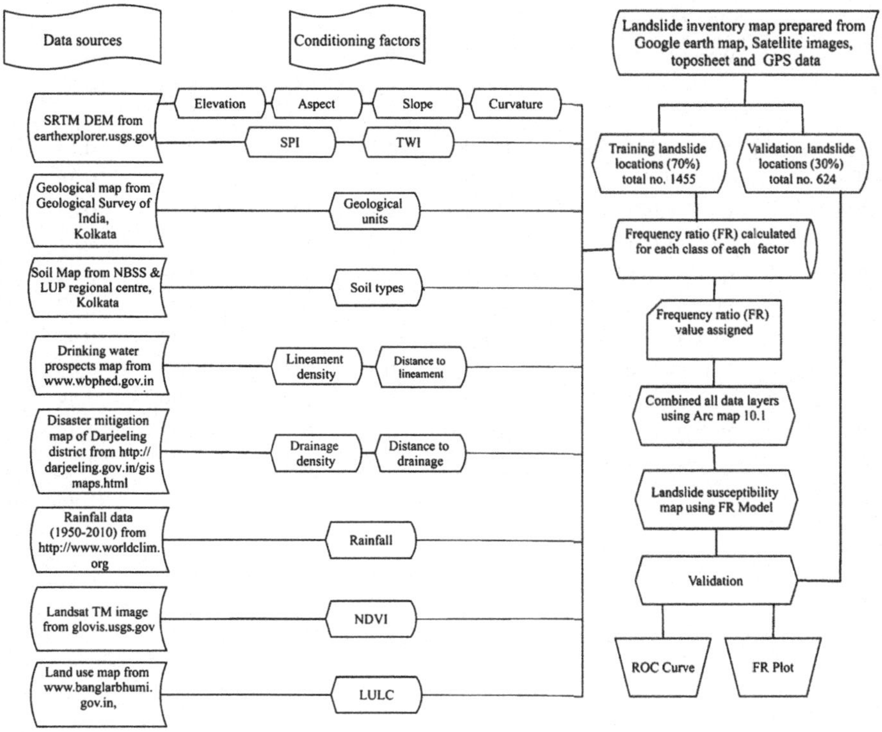

Fig. 3.1 Methodology of frequency ratio (FR) model

Table 3.1 Estimated frequency ratio value for each class of landslide conditioning factors

Conditioning factor	Class	Number of pixel	Total pixel (%)	Number of landslide occurrence pixels	Landslide occurrence pixels (%)	Frequency ratio value
Elevation (Meter)	130–470	409,786	15.89	465	9.53	0.60
	470–770	437,151	16.96	1519	31.11	1.83
	770–1100	437,588	16.97	1259	25.78	1.52
	1100–1300	246,682	9.57	445	9.11	0.95
	1300–1600	342,538	13.29	465	9.53	0.72
	1600–1900	260,070	10.09	178	3.65	0.36
	1900–2200	171,257	6.64	268	5.50	0.83
	2200–2600	142,581	5.53	207	4.25	0.77
	2600–3000	80,867	3.14	69	1.42	0.45
	3000–3600	49,605	1.92	6	0.12	0.06
Aspect	Flat (−1)	231	0.01	0	0.00	0.00
	North (0–22.5)	136,124	5.28	90	1.84	0.35
	Northeast (22.5–67.5)	296,730	11.51	706	14.47	1.26
	East (67.5–112.5)	395,973	15.36	890	18.22	1.19
	Southeast (112.5–157.5)	413,694	16.05	894	18.30	1.14
	South (157.5–202.5)	386,071	14.97	777	15.91	1.06
	Southwest (202.5–247.5)	299,310	11.61	844	17.29	1.49
	West (247.5–292.5)	266,605	10.34	396	8.11	0.78
	Northwest (292.5–337.5)	250,427	9.71	252	5.16	0.53
	North (337.5–360)	132,961	5.16	34	0.70	0.14
Slope (Degree)	0–4.339	187,019	7.25	134	2.74	0.38
	4.339–8.677	255,476	9.91	403	8.26	0.83
	8.677–12.515	343,632	13.33	590	12.08	0.91
	12.515–15.853	374,691	14.53	668	13.69	0.94
	15.853–19.024	427,124	16.57	746	15.27	0.92
	19.024–22.027	382,230	14.83	704	14.42	0.97
	22.027–25.031	278,609	10.81	573	11.73	1.09
	25.031–28.202	189,918	7.37	459	9.39	1.28
	28.202–32.373	107,393	4.17	501	10.27	2.47
	32.373–42.553	32,035	1.24	104	2.14	1.72

(continued)

Table 3.1 (continued)

Conditioning factor	Class	Number of pixel	Total pixel (%)	Number of landslide occurrence pixels	Landslide occurrence pixels (%)	Frequency ratio value
Curvature	−0.644 to −0.238	48,330	1.87	159	3.26	1.74
	−0.238 to −0.155	168,473	6.53	765	15.66	2.40
	−0.155 to −0.095	286,017	11.09	855	17.51	1.58
	−0.095 to −0.040	390,224	15.14	859	17.59	1.16
	−0.040 to 0.015	577,690	22.41	672	13.77	0.61
	0.015–0.064	406,708	15.78	548	11.23	0.71
	0.064–0.120	330,604	12.82	435	8.91	0.69
	0.120–0.185	234,280	9.09	330	6.75	0.74
	0.185–0.267	108,702	4.22	192	3.94	0.93
	0.267–0.756	27,099	1.05	67	1.38	1.31
Geology	Alluvium	104,604	4.06	27	0.55	0.14
	Baxa Series-Slates, Schists, Dolomite, Quartzites	2344	0.09	4	0.08	0.84
	Daling Series-Slates, Schists, Quartzites	712,203	27.62	1975	40.46	1.46
	Damuda (Gondwana)	35,988	1.40	116	2.38	1.71
	Darjeeling Gneiss	1,523,340	59.09	2631	53.90	0.91
	Nahan Group (Tertiary) Limestone	199,648	7.74	128	2.63	0.34
Soil	Tista River	14,861	0.58	0	0.00	0.00
	W001	261,470	10.14	492	10.08	0.99
	W002	1,434,692	55.65	2183	44.71	0.80
	W003	221,855	8.61	183	3.75	0.44
	W004	446,612	17.32	1967	40.30	2.33
	W006	75,842	2.94	1	0.02	0.01
	W007	85,141	3.30	17	0.34	0.10
	W008	35,338	1.37	39	0.80	0.58
	W009	2314	0.09	0	0.00	0.00
Drainage density (Km/Sq. Km)	0–0.694	243,293	9.44	266	5.44	0.58
	0.694–1.210	409,926	15.90	732	14.99	0.94
	1.210–1.620	469,031	18.19	879	18.00	0.99
	1.620–2.010	498,755	19.35	1202	24.62	1.27
	2.010–2.370	393,587	15.27	753	15.42	1.01
	2.370–2.730	272,462	10.57	665	13.62	1.29
	2.730–3.140	166,766	6.47	279	5.72	0.88
	3.140–3.680	76,489	2.97	83	1.69	0.57
	3.680–4.450	34,741	1.35	20	0.41	0.31
	4.450–6.560	13,076	0.51	4	0.08	0.16

(continued)

Table 3.1 (continued)

Conditioning factor	Class	Number of pixel	Total pixel (%)	Number of landslide occurrence pixels	Landslide occurrence pixels (%)	Frequency ratio value
Lineament density (Km/ Sq. Km)	0–0.056	1,309,367	50.79	2325	47.63	0.94
	0. 056–0.168	300,996	11.67	482	9.87	0.85
	0.168–0.287	230,519	8.94	577	11.81	1.32
	0.287–0.399	192,603	7.47	337	6.90	0.92
	0.399–0.505	162,361	6.30	342	7.00	1.11
	0.505–0.610	141,605	5.49	390	7.98	1.45
	0.610–0.716	102,252	3.97	248	5.09	1.28
	0.716–0.835	73,150	2.84	118	2.42	0.85
	0.835–0.966	48,815	1.89	53	1.08	0.57
	0.966–1.590	16,458	0.64	11	0.22	0.35
Distance from lineament (Meter)	0–20	23,691	0.92	42	0.86	0.93
	20–50	36,643	1.42	61	1.25	0.88
	50–100	63,438	2.46	119	2.43	0.99
	100–400	406,582	15.77	823	16.86	1.07
	400–1000	787,207	30.53	1426	29.22	0.96
	1000–2000	781,676	30.32	1448	29.66	0.98
	2000–3500	369,169	14.32	872	17.87	1.25
	3500–5000	95,146	3.69	91	1.87	0.51
	5000–6500	10,670	0.41	0	0.00	0.00
	6500–8500	3904	0.15	0	0.00	0.00
Distance to drainage (Meter)	0–10	92,856	3.60	210	4.31	1.20
	10–30	184,399	7.15	412	8.43	1.18
	30–70	356,416	13.82	754	15.44	1.12
	70–150	628,426	24.38	1291	26.45	1.09
	150–300	763,840	29.63	1599	32.76	1.11
	300–700	498,808	19.35	600	12.29	0.64
	700–1200	45,871	1.78	11	0.23	0.13
	1200–1700	5659	0.22	4	0.08	0.38
	1700–2300	1593	0.06	0	0.00	0.00
	2300–3000	259	0.01	0	0.00	0.00
SPI	−5.939 to −3.140	14,266	0.55	20	0.40	0.72
	−3.140 to −2.496	57,082	2.21	92	1.88	0.85
	−2.496 to −2.013	132,095	5.12	221	4.53	0.88
	−2.013 to −1.595	223,338	8.66	379	7.76	0.90
	−1.595 to −1.209	296,386	11.50	482	9.87	0.86
	−1.209 to −0.791	361,250	14.01	612	12.55	0.90
	−0.791 to −0.373	379,786	14.73	691	14.16	0.96
	−0.373 to 0.014	571,397	22.16	1040	21.29	0.96
	0.014–0.400	391,010	15.17	752	15.41	1.02
	0.400–2.266	151,516	5.88	593	12.15	2.07

(continued)

Table 3.1 (continued)

Conditioning factor	Class	Number of pixel	Total pixel (%)	Number of landslide occurrence pixels	Landslide occurrence pixels (%)	Frequency ratio value
TWI	−6907.755	231	0.01	0	0.00	0.00
	−6907.755 to −2849.717	848	0.03	0	0.00	0.00
	−2849.717 to −1226.502	2601	0.10	0	0.00	0.00
	−1226.502 to −520.757	6953	0.27	0	0.01	0.04
	−520.757 to −167.884	36,026	1.40	15	0.32	0.23
	−167.884 to −26.735	454,856	17.64	639	13.09	0.74
	−26.735 to 43.840	1,987,528	77.09	4150	85.00	1.10
	43.840–255.564	76,829	2.98	67	1.37	0.46
	255.564–714.298	10,964	0.43	10	0.21	0.49
	714.298–2090.502	1289	0.05	0	0.00	0.00
Mean annual rainfall (mm)	1076–1478	54,493	2.11	40	0.82	0.39
	1478–1798	70,178	2.72	43	0.89	0.33
	1798–2097	73,731	2.86	82	1.69	0.59
	2097–2355	88,710	3.44	287	5.89	1.71
	2355–2551	268,934	10.43	199	4.07	0.39
	2551–2705	551,234	21.38	909	18.63	0.87
	2705–2860	608,233	23.59	1341	27.48	1.16
	2860–3056	423,558	16.43	1165	23.87	1.45
	3056–3293	277,977	10.78	628	12.86	1.19
	3293–3705	161,077	6.25	187	3.82	0.61
NDVI	−0.220 to 0.037	34,132	1.32	324	6.64	5.03
	0.037–0.110	77,506	3.01	482	9.87	3.28
	0.110–0.166	167,231	6.49	520	10.65	1.64
	0.166–0.214	273,220	10.60	683	13.98	1.32
	0.214–0.263	381,618	14.80	759	15.54	1.05
	0.263–0.315	451,331	17.51	753	15.43	0.88
	0.315–0.368	428,060	16.60	587	12.02	0.72
	0.368–0.427	364,538	14.14	425	8.70	0.62
	0.427–0.493	257,616	9.99	245	5.02	0.50
	0.493–0.667	142,875	5.54	105	2.15	0.39

(continued)

Table 3.1 (continued)

Conditioning factor	Class	Number of pixel	Total pixel (%)	Number of landslide occurrence pixels	Landslide occurrence pixels (%)	Frequency ratio value
LULC	Social forestry	309,800	12.02	453	9.27	0.77
	Agricultural single crop	331,856	12.87	537	10.99	0.85
	Forest	1,430,826	55.50	2978	61.00	1.10
	Rural settlement	126,193	4.89	150	3.07	0.63
	Urban settlement	16,672	0.62	18	0.36	0.58
	Agricultural plantation (Tea)	337,588	13.09	514	10.53	0.80
	Wasteland dry/barren land	399	0.02	7	0.15	7.50
	Public utility and facility	3052	0.12	0	0.00	0.00
	Wastelands with scrub	21,740	0.84	225	4.61	5.49

3.2.2 *Information Value Model (IVM)*

According to Pereira et al. (2012), information value model (IVM) to evaluate the role of different combinations of landslide predisposing factors in the occurrence of shallow landslides in different parts of Northern Portugal. Kanungo et al. (2009), Champatiray (2000), Champatiray et al. (2007), and Arora et al. (2004) have opined that information value model (IVM) has proved useful method in determining the degree of influence of individual causative factor responsible for landslide occurrence. A modified form of pixel based information value model was also applied to assess landslide susceptibility mapping by Balsubramani and Kumaraswamy (2013). Some problems were ascertained in IV model, i.e. when no landslide exists in certain subclass, there is not any significance of that particular class. Researchers usually assigned 0 or "no pixel data" to those pixels (class), which would make the results much more exaggerated if a large number of those pixels existed, and since "0" value in the model means that the ratio of landslide pixels in subclass I is equal to the average ratio of the study area, and if no landslides existed, the results should approach infinitesimal. To overcome this problem, $N_{\mathrm{pix}(x_i)} = 0, I(H, x_i)$ was not calculated and was qualitatively determined as the lower information value considering the data set of predisposing variables, which could avoid the problem of high exaggeration, while the results could not exactly show the information value of this area by Oliveira et al. (2015). It showed that MIV model can tackle the problem of "no pixel data in subclass" well, generate the true information value, and show real running trend, which performs well in showing the relationship between predisposing factors and landslide occurrence and can be used for preliminary landslide susceptibility assessment in the study area. In the present study, modified information value model was applied to assess slope instability of Darjeeling Himalaya.

Spatial distribution of landslide susceptibility was analysed with reasonable accuracy using information value model on GIS platform by Vijith et al. (2009), Sarkar et al. (2006), Zezere (2002), Wang and Sassa (2005), Sharma et al. (2009), Akbar and Ha (2011), and Pereira et al. (2012). The information value (IV) for each subclass of the factors was calculated with the help of following equation (eq. 3.3).

$$I(H,\mathrm{x_i}) = \mathrm{In}\frac{N_{\mathrm{pix(x_i)}} / N_{\mathrm{pix(N_i)}}}{\Sigma N_{\mathrm{pix(x_i)}} / \Sigma N_{\mathrm{pix(N_i)}}} \quad \text{(Eq. 3.3)}$$

where

$I(H, \mathrm{x_i})$ = the information value of subclass i of a predisposing factor
$N_{\mathrm{pix(x_i)}}$ = the number of landslide pixels in subclass i
$N_{\mathrm{pix(N_i)}}$ = the total pixel number of subclass i
$\Sigma N_{\mathrm{pix(x_i)}}$ = the total landslide pixel number of in the study area
$\Sigma N_{\mathrm{pix(N_i)}}$ = the total pixel number of the study area

Therefore, the total information value I_{total} for each pixel was calculated by summing up all the information values of each factor layer (eq. 3.4).

$$I_{\mathrm{total}} \sum_{n}^{i=1} I(H,\mathrm{x_i}) \quad \text{(Eq. 3.4)}$$

where n = is the number of layers of the predisposing factors.

3.2.3 *Modified Information Value Model (MIVM)*

MIV model can be applied for the landslide susceptibility zonation mapping (Wang et al. 2015; Mandal and Mandal 2017) to avoid the problem ("no pixel data" for a certain subclass) and to quantitatively express the results, Equation (3.3) in IV model was modified and MIV model applied in the present work (eq. 3.5).

$$I(H,\mathrm{x_i}) = \log 2\frac{N_{\mathrm{pix(x_i)}} / N_{\mathrm{pix(N_i)}}}{\Sigma N_{\mathrm{pix(x_i)}} / \Sigma N_{\mathrm{pix(N_i)}}} + 1 \quad \text{(Eq. 3.5)}$$

where when no landslides existed in a certain subclass, $I(H, x_i)$ equals "0", which would be the smallest value; when $I(H, x_i)$ is "1", it means that the ratio of landslide pixels in subclass i equals to the average of the study area; and when the value is larger than "1", more landslides pixels in the subclass lie than those in the study area, and the larger the value, the higher landslides ratio in the subclass.

This model was used to determine the modified information values for each class of the landslide causative factors and the total information value of each cell of the area (Table 3.2). The greater the total information value, the greater is the probability of landslide susceptibility.

3.2.4 *Models Validation*

To validate the prepared landslide susceptibility maps derived by using frequency ratio model and modified information value model, landslide locations map was made in consultation with topographical map, Google earth image, GPS points, and historical landslide records (Fig. 3.2). Researchers made landslide inventory map (Fig. 3.3) with the help of aerial photographs and satellite imageries (Lee and Pradhan 2006a, b; Pradhan and Lee 2009; Pradhan 2010; Choi et al. 2011; Umar et al. 2014). But in the present study, very small to large polygon coverage landslide locations were being considered using Survey of India Topographical map and Google earth historical imageries (2015–2016). Not only that an intensive field investigation with GPS was conducted to evaluate the landslide inventory map of Darjeeling Himalaya. Finally, the prepared landslide inventory map was digitized and converted into raster value domain in Arc Map (10.1) environment. The prepared landslide susceptibility maps based on *frequency ratio model* and *modified information value model* were being validated incorporating area under curve of ROC curve. To develop the ROC curve, a total of 2079 data sets were used out of which 1455 was training data set and 624 was validation data set.

3.3 Result and Discussion

3.3.1 *Frequency Ratio (FR) Model and Landslide Susceptibility*

Darjeeling Himalaya was classified into five landslide susceptibility zones, i.e. very low, low, moderate, high, and very high (Fig. 3.4). The study revealed that 24.60% area of Darjeeling Himalaya is registered with moderate landslide susceptibility. More than 50% area of Darjeeling Himalaya is characterized by moderate to high

Table 3.2 Class wise estimated modified information value

Conditioning factor	Class	No. of landslide occurrence pixels($N_{pix(x_i)}$)	No. of pixel ($N_{pix(N_i)}$)	$N_{pix(x_i)} / N_{pix(N_i)}$	The total landslide pixel number in the study area ($\sum N_{pix(x_i)}$)	The total pixel number in the study area ($\sum N_{pix(N_i)}$)	$\sum N_{pix(x_i)} / \sum N_{pix(N_i)}$	Information value (IV)	Modified information value (MIV)
Elevation (m)	130–470	465	409,786	0.00113474	4882	2,578,126	0.00189362	−0.51	−1.04
	470–770	1519	437,151	0.00347477	4882	2,578,126	0.00189362	0.61	0.68
	770–1100	1259	437,588	0.00287714	4882	2,578,126	0.00189362	0.42	0.50
	1100–1300	445	246,682	0.00180394	4882	2,578,126	0.00189362	−0.05	−0.07
	1300–1600	465	342,538	0.00135751	4882	2,578,126	0.00189362	−0.33	−0.58
	1600–1900	178	260,070	0.00068443	4882	2,578,126	0.00189362	−0.69	−1.69
	1900–2200	268	171,257	0.0015649	4882	2,578,126	0.00189362	−0.19	−0.31
	2200–2600	207	142,581	0.00145181	4882	2,578,126	0.00189362	−0.27	−0.45
	2600–3000	69	80,867	0.00085325	4882	2,578,126	0.00189362	−0.80	−2.30
	3000–3600	6	49,605	0.00012096	4882	2,578,126	0.00189362	−0.95	−4.32

(continued)

Table 3.2 (continued)

Conditioning factor	Class	No. of landslide occurrence pixels($N_{pix(x_i)}$)	No. of pixel ($N_{pix(N_i)}$)	$N_{pix(x_i)} / N_{pix(N_i)}$	The total landslide pixel number in the study area ($\sum N_{pix(x_i)}$)	The total pixel number in the study area ($\sum N_{pix(N_i)}$)	$\sum N_{pix(x_i)} / \sum N_{pix(N_i)}$	Information value (IV)	Modified information value (MIV)
Aspect	Flat (−1)	0	231	0	4882	2,578,126	0.00189362	0.00	0.00
	North (0–22.5)	90	136,124	0.00066116	4882	2,578,126	0.00189362	−0.70	−1.74
	Northeast (22.5–67.5)	706	296,730	0.00237927	4882	2,578,126	0.00189362	0.23	0.30
	East (67.5–112.5)	890	395,973	0.00224763	4882	2,578,126	0.00189362	0.17	0.23
	Southeast (112.5–157.5)	894	413,694	0.00216102	4882	2,578,126	0.00189362	0.13	0.18
	South (157.5–202.5)	777	386,071	0.00201258	4882	2,578,126	0.00189362	0.06	0.09
	Southwest (202.5–247.5)	844	299,310	0.00281982	4882	2,578,126	0.00189362	0.40	0.48
	West (247.5–292.5)	396	266,605	0.00148534	4882	2,578,126	0.00189362	−0.24	−0.40
	Northwest (292.5–337.5)	252	250,427	0.00100628	4882	2,578,126	0.00189362	−0.63	−1.44
	North (337.5–360)	34	132,961	0.00025571	4882	2,578,126	0.00189362	−0.89	−3.18

Slope (degree)	0–4.339	134	187,019	0.0007165	4882	2,578,126	0.00189362	−0.97	−5.15
	4.339–8.677	403	255,476	0.00157745	4882	2,578,126	0.00189362	−0.18	−0.29
	8.677–12.515	590	343,632	0.00171695	4882	2,578,126	0.00189362	−0.10	−0.15
	12.515–15.853	668	374,691	0.0017828	4882	2,578,126	0.00189362	−0.06	−0.09
	15.853–19.024	746	427,124	0.00174657	4882	2,578,126	0.00189362	−0.08	−0.12
	19.024–22.027	704	382,230	0.00184182	4882	2,578,126	0.00189362	−0.03	−0.04
	22.027–25.031	573	278,609	0.00205665	4882	2,578,126	0.00189362	0.08	0.11
	25.031–28.202	459	189,918	0.00241683	4882	2,578,126	0.00189362	0.24	0.31
	28.202–32.373	501	107,393	0.00466511	4882	2,578,126	0.00189362	0.90	0.93
	32.373–42.553	104	32,035	0.00324645	4882	2,578,126	0.00189362	0.54	0.62
Curvature	−0.644 to −0.238	159	48,330	0.00328988	4882	2,578,126	0.00189362	0.55	0.63
	−0.238 to −0.155	765	168,473	0.00454079	4882	2,578,126	0.00189362	0.87	0.91
	−0.0155 to −0.095	855	286,017	0.00298933	4882	2,578,126	0.00189362	0.46	0.54
	−0.095 to −0.040	859	390,224	0.0022013	4882	2,578,126	0.00189362	0.15	0.20
	−0.040 to −0.015	672	577,690	0.00116325	4882	2,578,126	0.00189362	−0.49	−0.96
	0.015–0.064	548	406,708	0.0013474	4882	2,578,126	0.00189362	−0.34	−0.60
	0.064–0.120	435	330,604	0.00131577	4882	2,578,126	0.00189362	−0.36	−0.65
	0.120–0.185	330	234,280	0.00140857	4882	2,578,126	0.00189362	−0.30	−0.51
	0.185–0.267	192	108,702	0.0017663	4882	2,578,126	0.00189362	−0.07	−0.10
	0.267–0.756	67	27,099	0.00247242	4882	2,578,126	0.00189362	0.27	0.34

(continued)

Table 3.2 (continued)

Conditioning factor	Class	No. of landslide occurrence pixels($N_{pix(x_i)}$)	No. of pixel ($N_{pix(N_i)}$)	$N_{pix(x_i)} / N_{pix(N_i)}$	The total landslide pixel number in the study area ($\sum N_{pix(x_i)}$)	The total pixel number in the study area ($\sum N_{pix(N_i)}$)	$\sum N_{pix(x_i)} / \sum N_{pix(N_i)}$	Information value (IV)	Modified information value (MIV)
Geology	Alluvium	27	104,604	0.00025812	4882	2,578,126	0.00189362	−0.88	−3.06
	Baxa Series-Slates, Schists, Dolomite, Quartzites	4	2344	0.00170648	4882	2,578,126	0.00189362	−0.10	−0.16
	Daling Series-Slates, Schists, Quartzites	1975	712,203	0.00277309	4882	2,578,126	0.00189362	0.38	0.47
	Damuda (Gondwana)	116	35,988	0.0032233	4882	2,578,126	0.00189362	0.53	0.62
	Darjeeling Gneiss	2631	1,523,340	0.00172713	4882	2,578,126	0.00189362	−0.09	−0.14
	Nahan Group (Tertiary) Limestone	128	199,648	0.00064113	4882	2,578,126	0.00189362	−1.08	#NUM!

Gravity anomaly (mGal)	−31.584	0	9009	0	4882	2,578,126	0.00189362	0.00	0.00
	−16.069	79	154,668	0.00051077	4882	2,578,126	0.00189362	−0.77	−2.12
	−8.311	283	287,775	0.00098341	4882	2,578,126	0.00189362	−0.66	−1.54
	−7.204	1283	441,508	0.00290595	4882	2,578,126	0.00189362	0.43	0.51
	−7.203	1497	554,694	0.00269879	4882	2,578,126	0.00189362	0.35	0.44
	−7.758	788	482,165	0.0016343	4882	2,578,126	0.00189362	−0.15	−0.23
	−9.42	732	327,022	0.00223838	4882	2,578,126	0.00189362	0.17	0.22
	−12.744	182	164,458	0.00110667	4882	2,578,126	0.00189362	−0.54	−1.11
	−15.512	33	112,345	0.00029374	4882	2,578,126	0.00189362	−0.86	−2.84
	−25.492	5	44,483	0.0001124	4882	2,578,126	0.00189362	−0.97	−5.06
Soil	Tista River	0	14,861	0	4882	2,578,126	0.00189362	0.00	0.00
	W001	492	261,470	0.00188167	4882	2,578,126	0.00189362	−0.01	−0.01
	W002	2183	1,434,692	0.00152158	4882	2,578,126	0.00189362	−0.22	−0.36
	W003	183	221,855	0.00082486	4882	2,578,126	0.00189362	−0.83	−2.57
	W004	1967	446,612	0.00440427	4882	2,578,126	0.00189362	0.84	0.88
	W006	1	75,842	1.3185E-05	4882	2,578,126	0.00189362	0.99	0.99
	W007	17	85,141	0.00019967	4882	2,578,126	0.00189362	0.93	0.95
	W008	39	35,338	0.00110363	4882	2,578,126	0.00189362	−0.54	−1.12
	W009	0	2314	0	4882	2,578,126	0.00189362	0.00	0.00

(continued)

Table 3.2 (continued)

Conditioning factor	Class	No. of landslide occurrence pixels($N_{pix(x_i)}$)	No. of pixel ($N_{pix(N_i)}$)	$N_{pix(x_i)} / N_{pix(N_i)}$	The total landslide pixel number in the study area ($\sum N_{pix(x_i)}$)	The total pixel number in the study area ($\sum N_{pix(N_i)}$)	$\sum N_{pix(x_i)} / \sum N_{pix(N_i)}$	Information value (IV)	Modified information value (MIV)
Drainage density (km/ sq. km)	0–0.694	266	243,293	0.00109333	4882	2,578,126	0.00189362	−0.55	−1.15
	0.694–1.210	732	409,926	0.00178569	4882	2,578,126	0.00189362	−0.06	−0.09
	1.210–1.620	879	469,031	0.00187408	4882	2,578,126	0.00189362	−0.01	−0.02
	1.620–2.010	1202	498,755	0.00241	4882	2,578,126	0.00189362	0.24	0.31
	2.010–2.370	753	393,587	0.00191317	4882	2,578,126	0.00189362	0.01	0.01
	2.370–2.730	665	272,462	0.00244071	4882	2,578,126	0.00189362	0.25	0.33
	2.730–3.140	279	166,766	0.001673	4882	2,578,126	0.00189362	−0.12	−0.19
	3.140–3.680	83	76,489	0.00108512	4882	2,578,126	0.00189362	−0.56	−1.17
	3.680–4.450	20	34,741	0.00057569	4882	2,578,126	0.00189362	−0.76	−2.06
	0–0.694	4	13,076	0.0003059	4882	2,578,126	0.00189362	−0.84	−2.64
Lineament density (km/ sq. km)	0–0.056	2325	1,309,367	0.00177567	4882	2,578,126	0.00189362	−0.06	−0.10
	0. 056–0.168	482	300,996	0.00160135	4882	2,578,126	0.00189362	−0.17	−0.26
	0.168–0.287	577	230,519	0.00250305	4882	2,578,126	0.00189362	0.28	0.36
	0.287–0.399	337	19,2603	0.00174971	4882	2,578,126	0.00189362	−0.08	−0.12
	0.399–0.505	342	162,361	0.00210642	4882	2,578,126	0.00189362	0.11	0.15
	0.505–0.610	390	141,605	0.00275414	4882	2,578,126	0.00189362	0.37	0.46
	0.610–0.716	248	102,252	0.00242538	4882	2,578,126	0.00189362	0.25	0.32
	0.716–0.835	118	73,150	0.00161312	4882	2,578,126	0.00189362	−0.16	−0.25
	0.835–0.966	53	48,815	0.00108573	4882	2,578,126	0.00189362	−0.56	−1.17
	0.966 – 1.590	11	16,458	0.00066837	4882	2,578,126	0.00189362	−0.82	−2.47

Distance from lineament (m)	0–20	42	23,691	0.00177283	4882	2,578,126	0.00189362	−0.07	−0.10
	20–50	61	36,643	0.00166471	4882	2,578,126	0.00189362	−0.13	−0.20
	50–100	119	63,438	0.00187585	4882	2,578,126	0.00189362	−0.01	−0.01
	100–400	823	406,582	0.00202419	4882	2,578,126	0.00189362	0.07	0.09
	400–1000	1426	787,207	0.00181147	4882	2,578,126	0.00189362	−0.04	−0.07
	1000–2000	1448	781,676	0.00185243	4882	2,578,126	0.00189362	−0.02	−0.03
	2000–3500	872	369,169	0.00236206	4882	2,578,126	0.00189362	0.22	0.29
	3500–5000	91	95,146	0.00095642	4882	2,578,126	0.00189362	−0.68	−1.66
	5000–6500	0	10,670	0	4882	2,578,126	0.00189362	0.00	0.00
	6500–8500	0	3904	0	4882	2,578,126	0.00189362	0.00	0.00
Distance to drainage (m)	0–10	210	92,856	0.00226157	4882	2,578,126	0.00189362	0.18	0.24
	10–30	412	184,399	0.00223429	4882	2,578,126	0.00189362	0.17	0.22
	30–70	754	356,416	0.00211551	4882	2,578,126	0.00189362	0.11	0.15
	70–150	1291	628,426	0.00205434	4882	2,578,126	0.00189362	0.08	0.11
	150–300	1599	763,840	0.00209337	4882	2,578,126	0.00189362	0.10	0.14
	300–700	600	498,808	0.00120287	4882	2,578,126	0.00189362	−0.45	−0.87
	700–1200	11	45,871	0.0002398	4882	2,578,126	0.00189362	−0.91	−3.47
	1200–1700	4	5659	0.00070684	4882	2,578,126	0.00189362	−0.99	−6.10
	1700–2300	0	1593	0	4882	2,578,126	0.00189362	0.00	0.00
	2300–3000	0	259	0	4882	2,578,126	0.00189362	0.00	0.00

(continued)

Table 3.2 (continued)

Conditioning factor	Class	No. of landslide occurrence pixels($N_{pix(x_i)}$)	No. of pixel ($N_{pix(N_i)}$)	$N_{pix(x_i)} / N_{pix(N_i)}$	The total landslide pixel number in the study area ($\sum N_{pix(x_i)}$)	The total pixel number in the study area ($\sum N_{pix(N_i)}$)	$\sum N_{pix(x_i)} / \sum N_{pix(N_i)}$	Information value (IV)	Modified information value (MIV)
SPI	–5.939 to –3.140	20	14,266	0.00140193	4882	2,578,126	0.00189362	−0.30	−0.52
	–3.140 to –2.496	92	57,082	0.00161172	4882	2,578,126	0.00189362	−0.16	−0.25
	–2.496 to –2.013	221	132,095	0.00167304	4882	2,578,126	0.00189362	−0.12	−0.19
	–2.013 to –1.595	379	223,338	0.00169698	4882	2,578,126	0.00189362	−0.11	−0.17
	–1.595 to –1.209	482	296,386	0.00162626	4882	2,578,126	0.00189362	−0.15	−0.24
	–1.209 to –0.791	612	361,250	0.00169412	4882	2,578,126	0.00189362	−0.11	−0.17
	–0.791 to –0.373	691	379,786	0.00181945	4882	2,578,126	0.00189362	−0.04	−0.06
	–0.373 to –0.014	1040	571,397	0.0018201	4882	2,578,126	0.00189362	−0.04	−0.06
	0.014–0.400	752	391,010	0.00192322	4882	2,578,126	0.00189362	0.02	0.02
	0.400–2.266	593	151,516	0.00391378	4882	2,578,126	0.00189362	0.73	0.79

TWI	−6907.76	0	231	0	4882	2,578,126	0.00189362	0.00	0.00
	−6907.755 to −2849.717	0	848	0	4882	2,578,126	0.00189362	0.00	0.00
	−2849.717 to −1226.502	0	2601	0	4882	2,578,126	0.00189362	0.00	0.00
	−1226.502 to −520.757	0	6953	0	4882	2,578,126	0.00189362	0.00	0.00
	−520.757 to −167.884	15	36,026	0.00041637	4882	2,578,126	0.00189362	−0.80	−2.32
	−167.884 to −26.735	639	454,856	0.00140484	4882	2,578,126	0.00189362	−0.30	−0.51
	−26.735 to −43.840	4150	1,987,528	0.00208802	4882	2,578,126	0.00189362	0.10	0.13
	43.840–255.564	67	76,829	0.00087207	4882	2,578,126	0.00189362	−0.78	−2.15
	255.564–714.298	10	10,964	0.00091208	4882	2,578,126	0.00189362	−0.73	−1.89
	714.298–2090.502	0	1289	0	4882	2,578,126	0.00189362	0.00	0.00

(continued)

Table 3.2 (continued)

Conditioning factor	Class	No. of landslide occurrence pixels($N_{pix(x_i)}$)	No. of pixel ($N_{pix(N_i)}$)	$N_{pix(x_i)} / N_{pix(N_i)}$	The total landslide pixel number in the study area ($\sum N_{pix(x_i)}$)	The total pixel number in the study area ($\sum N_{pix(N_i)}$)	$\sum N_{pix(x_i)} / \sum N_{pix(N_i)}$	Information value (IV)	Modified information value (MIV)
Mean annual rainfall (mm)	1076–1478	40	54,493	0.00073404	4882	2,578,126	0.00189362	−0.95	−4.26
	1478–1798	43	70,178	0.00061273	4882	2,578,126	0.00189362	−0.72	−1.84
	1798–2097	82	73,731	0.00111215	4882	2,578,126	0.00189362	−0.53	−1.10
	2097–2355	287	88,710	0.00323526	4882	2,578,126	0.00189362	0.54	0.62
	2355–2551	199	268,934	0.00073996	4882	2,578,126	0.00189362	−0.94	−4.05
	2551–2705	909	551,234	0.00164903	4882	2,578,126	0.00189362	−0.14	−0.21
	2705–2860	1341	608,233	0.00220475	4882	2,578,126	0.00189362	0.15	0.20
	2860–3056	1165	423,558	0.00275051	4882	2,578,126	0.00189362	0.37	0.46
	3056–3293	628	27,7977	0.00225918	4882	2,578,126	0.00189362	0.18	0.23
	1076–1478	187	161,077	0.00116094	4882	2,578,126	0.00189362	−0.49	−0.97
NDVI	−0.220 to 0.037	324	34,132	0.00949256	4882	2,578,126	0.00189362	1.61	1.39
	0.037–0.110	482	77,506	0.00621887	4882	2,578,126	0.00189362	1.19	1.13
	0.110–0.166	520	16,7231	0.00310947	4882	2,578,126	0.00189362	0.50	0.58
	0.166–0.214	683	273,220	0.00249982	4882	2,578,126	0.00189362	0.28	0.35
	0.214–0.263	759	381,618	0.0019889	4882	2,578,126	0.00189362	0.05	0.07
	0.263–0.315	753	451,331	0.0016684	4882	2,578,126	0.00189362	−0.13	−0.20
	0.315–0.368	587	428,060	0.0013713	4882	2,578,126	0.00189362	−0.32	−0.56
	0.368–0.427	425	364,538	0.00116586	4882	2,578,126	0.00189362	−0.49	−0.96
	0.427–0.493	245	257,616	0.00095103	4882	2,578,126	0.00189362	−0.69	−1.68
	0.493–0.667	105	142,875	0.00073491	4882	2,578,126	0.00189362	−0.95	−4.22

LULC	Social forestry	453	309,800	0.00146223	4882	2,578,126	0.00189362	−0.26	−0.43
	Agricultural single Crop	537	331,856	0.00161817	4882	2,578,126	0.00189362	−0.16	−0.25
	Forest	2978	1,430,826	0.00208132	4882	2,578,126	0.00189362	0.09	0.13
	Rural settlement	150	126,193	0.00118866	4882	2,578,126	0.00189362	−0.47	−0.90
	Urban settlement	18	16,672	0.00107965	4882	2,578,126	0.00189362	−0.56	−1.19
	Agricultural plantation (Tea)	514	337,588	0.00152257	4882	2,578,126	0.00189362	−0.22	−0.35
	Wasteland dry/barren land	7	399	0.01754386	4882	2,578,126	0.00189362	2.23	1.69
	Public utility and facility	0	3052	0	4882	2,578,126	0.00189362	0.00	0.00
	Wastelands with scrub	225	21,740	0.01034959	4882	2,578,126	0.00189362	1.70	1.43

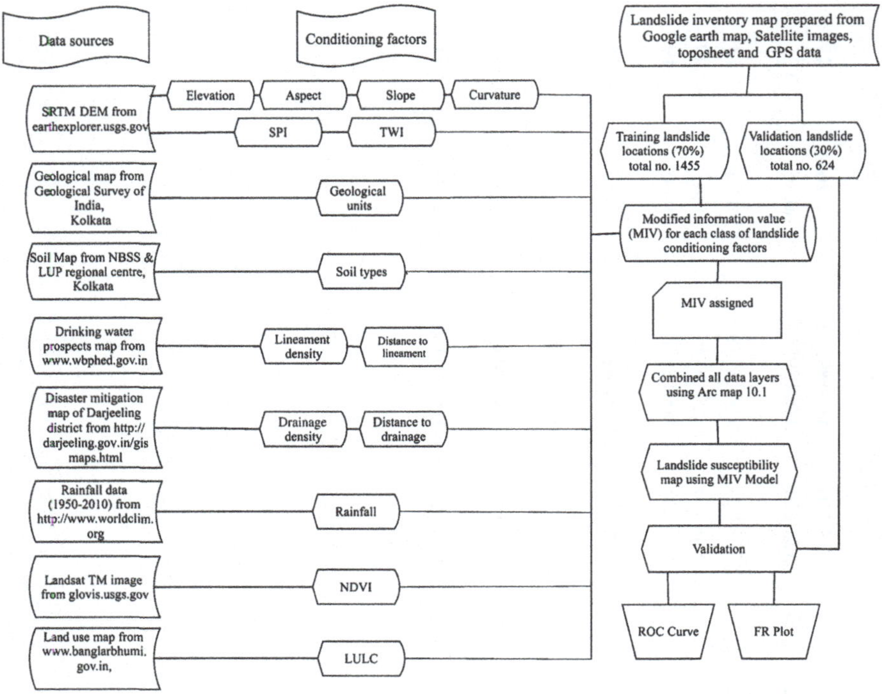

Fig. 3.2 Methodology of modified information value (MIV) model

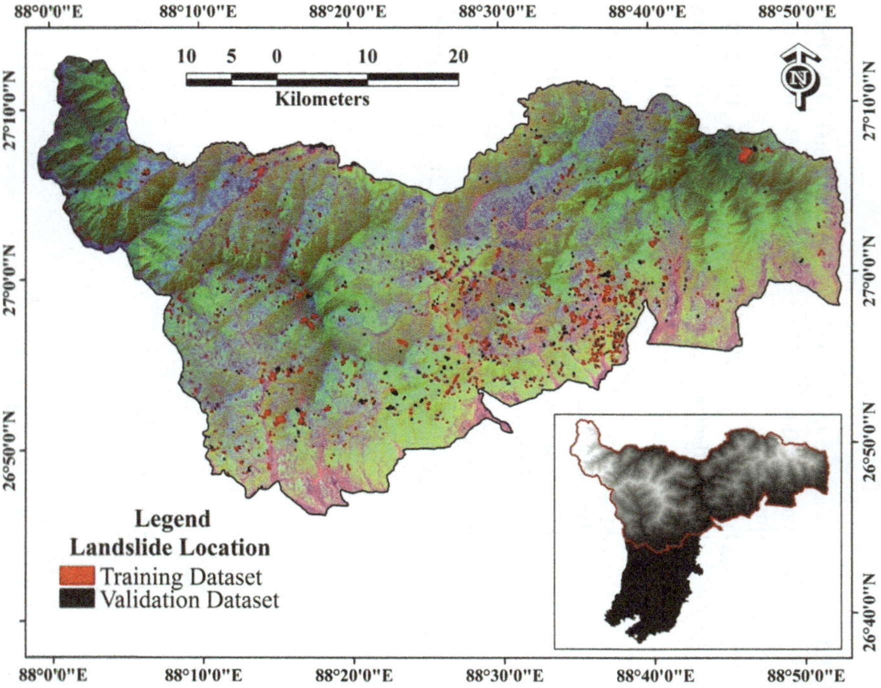

Fig. 3.3 Landslide inventory map of Darjeeling Himalaya

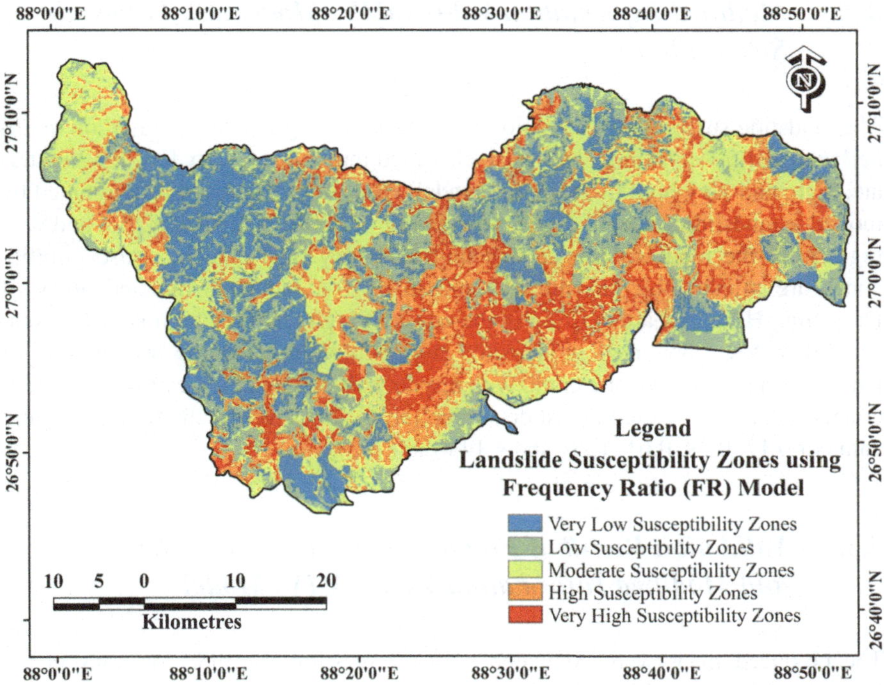

Fig. 3.4 Landslide susceptibility map using frequency ratio model (FRM)

Table 3.3 Area and frequency ratio value of different susceptibility classes using FR and MIV models

	Total pixels		% of Total pixels		Landslide pixels		% of landslide pixels		Frequency ratio	
Landslide susceptibility zones	Using FR model	Using MIV model	Using FR model	Using MIV model	Using FR model	Using MIV model	Using FR model	Using MIV model	Using FR model	Using MIV model
Very low	462,273	248,243	17.93	9.63	249	82	5.09	1.68	0.28	0.17
Low	599,935	551,662	23.27	21.40	695	304	14.24	6.23	0.61	0.29
Moderate	634,341	888,142	24.60	34.45	774	1033	15.85	21.16	0.64	0.61
High	599,588	644,030	23.26	24.98	1396	1534	28.60	31.43	1.23	1.26
Very high	281,988	246,049	10.94	9.54	1768	1928	36.22	39.49	3.31	4.14

landslide susceptibility. It was observed that high susceptible area and very high landslide susceptible areas are associated with 28.60% and 36.22% landslide affected areas, respectively (Table 3.3). The estimated frequency ratio value of each landslide susceptibility class also denotes that the landslide probability is very high at the places where landslide affected area are maximum and where landslide susceptibility is also high.

3.3.2 Modified Information Value (MIV) Model and Landslide Susceptibility

The landslide susceptibility zonation map of Darjeeling Himalaya based on *modified information value model* was also classified into five, i.e. very low, low, moderate, high, and very high (3.5). In this model, Darjeeling Himalaya is dominated by moderate landslide susceptibility which is followed by high, low, very low, and very high. Both the models depicted that high and very high landslide susceptibility zones are affected by landslides. More than 70% landslide affected areas of Darjeeling Himalaya are existing in the zones of high and very high landslide susceptibility. Very low and low landslide susceptible places having less number of landslides and low frequency ratio with respect to both models. Very low, low, moderate, high, and very high landslide susceptibility zones having the frequency ratio value of 0.17, 0.29, 0.61, 1.26, and 4.14 (Table 3.3).

3.3.3 Validation Result of Frequency Ratio (FR) Model and Modified Information Value (MIV) Model

The produced maps were compared with the existing landslide locations in the Darjeeling Himalaya (Fig. 3.5). For validation, success rate curve was made by comparing the existing landslide locations with the landslide susceptibility map (Bui et al.,

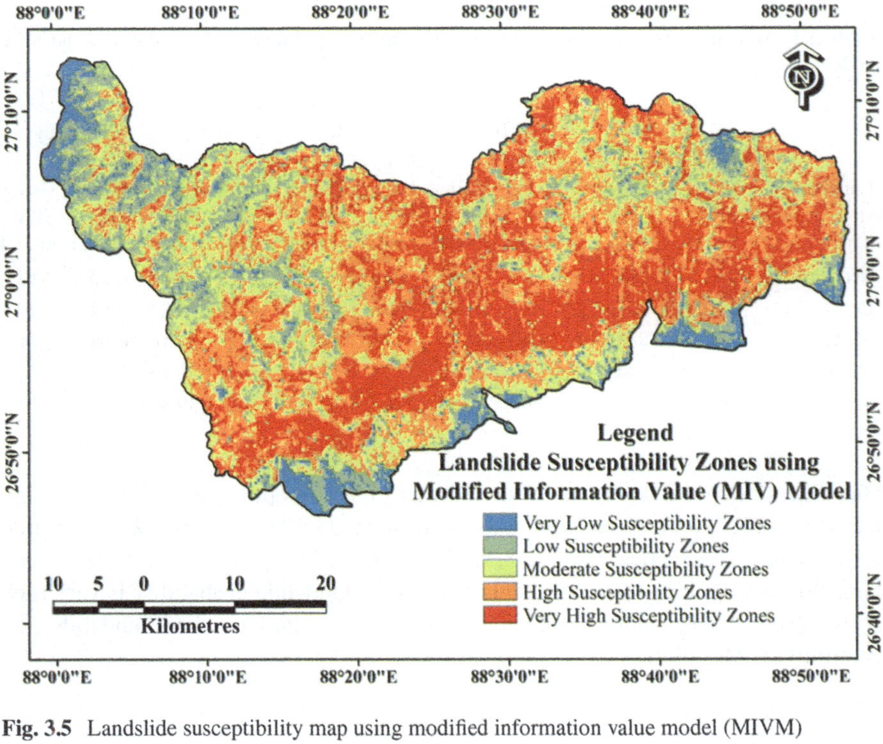

Fig. 3.5 Landslide susceptibility map using modified information value model (MIVM)

2011). The success rate results were obtained by using the training data set of 624. The success rate curve for frequency ratio model shows that the area under the curve (AUC) is 0.746 and modified information value model is 0.769, which corresponds to the prediction accuracy of 74.60% and 76.90%, respectively (Fig. 3.6). Frequency ratio plot of both the models suggested that, the chance of landslide occurrences was gradually decreasing from very high to very low susceptibility zones (Fig. 3.7). This was an indication of ideal landslide susceptibility index maps.

Fig. 3.6 ROC of Frequency ratio model and modified information value model

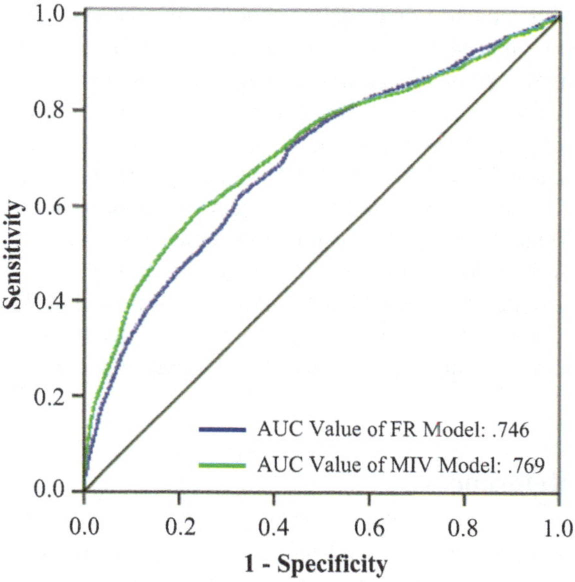

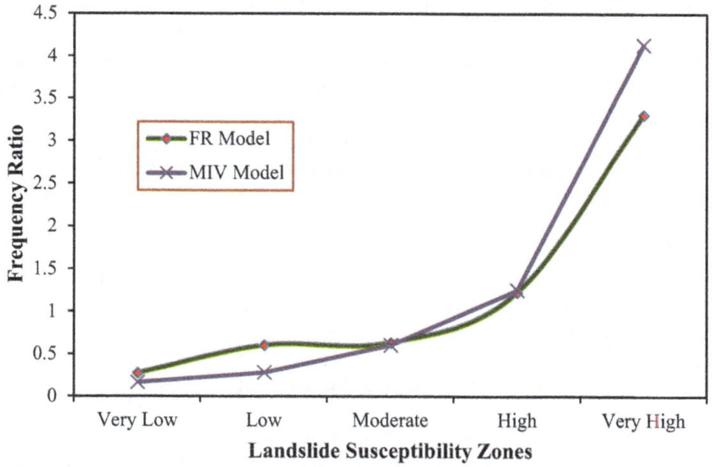

Fig. 3.7 Frequency ratio plot of different landslide susceptibility zones using FR and MIV models

3.4 Conclusion

Landslide susceptibility of the mountain environment of Darjeeling Himalaya should be analysed for identifying probable landslide susceptible areas as well as to mitigate damages caused by landslide phenomena. The prepared landslide susceptibility map of Darjeeling Himalaya using frequency ratio model and modified information value model showed very low, low, moderate, high, and very high landslide susceptibility zones of Darjeeling Himalaya. Both the models established the relationship between landslide conditioning factors and landslide inventory or landslide distribution map. By integrating several factors, i.e. elevation, slope aspect, slope angle, slope curvature, geology, soil, lineament density, distance to lineament, drainage density, distance to drainage, stream power index (SPI), topographic wetted index (TWI), rainfall, normalized differential vegetation index (NDVI) and land use and land cover (LULC) frequency ratio value, and modified information value were being extracted for each class to understand the relative importance of individual factor and corresponding classes. The prepared landslide susceptibility maps, derived AUC values of frequency ratio model, and modified frequency ratio model revealed the prediction accuracy of 74.60% and 76.90%, respectively. The study revealed that the modified information value model is suitable to prepare landslide susceptibility map of Darjeeling Himalaya in comparison to frequency ratio model.

References

Akbar T, Ha S (2011) Landslide hazard zoning along Himalaya KaghanValley of Pakistan-by integration of GPS, GIS, and remote sensingtechnology. *Landslides, 8*(4), 527–540.

Arora, M., Das Gupta, A., & Gupta, R. (2004). An artificial neural network approach for landslide hazard zonation in the Bhagirathi (Ganga) Valley, Himalaya. *International Journal of Remote Sensing, 25*, 559–572. https://doi.org/10.1080/0143116031000156819.

Avinash, K. G., & Ashamanjari, K. G. (2010). A GIS and frequency ratio based landslide susceptibility mapping: Aghnashini river catchment, Uttara Kannada, India. *International Journal of Geomatics and Geosciences, 1*(3), 343–354.

Atkinson, P. M., & Massari, R. (1998). Generalized linear modelling of susceptibility to landsliding in the central Apennines, Italy. *Computers & Geosciences, 24*, 373–385.

Bagherzadeh, A., & Mansouri Daneshvar, M. R. (2012). Mapping of landslide hazard zonationusing GIS at Golestan watershed, northeast of Iran. *Arabian Journal of Geosciences, 6*, 3377–3388.

Balsubramani, K., & Kumaraswamy, K. (2013). Application of geospatial technology andinformation value technique in landslide hazard zonation mapping: A case study of Giri Valley, Himachal Pradesh. *Disaster Advances, 6*, 38–47.

Bui, D. T., Lofman, O., Revhaug, I., & Dick, O. (2011). Landslide susceptibility analysis in the Hoa Binh province of Vietnam using statistical index and logistic regression. *Natural Hazards, 59*(3), 1413–1444.

Caiyan WU and Jianping Q (2009) Relationship between landslides and lithology in the Three Gorges Reservoir area based on GIS AND Information Value Model. Higher Education Press and Springer New York *4*(2), 165–170

Champatiray, P. (2000). Perationalization of costeffective methodology for landslide hazard zonation using RS and GIS: IIRS initiative. In P. Roy, C. Van Westen, V. Jha, & R. Lakhera (Eds.), *Natural disasters and their mitigation; remote sensing and geographical information system perspectives* (pp. 95–101). Dehradun, India: Indian Institute of Remote Sensing.

Champatiray, P., Dimri, S., Lakhera, R., & Sati, S. (2007). Fuzzy based methods for landslide hazard assessment in active seismic zone of Himalaya. *Landslides, 4*, 101–110. https://doi.org/10.1007/s1034600600686.

Choi, J., Oh, H. J., Lee, H. J., Lee, C., & Lee, S. (2011). Combining landslide susceptibility maps obtained from frequency ratio, logistic regression, and artificial neural network models using aster images and GIS. *Engineering Geology, 124*, 12–23.

Donati, L., & Turrini, M. C. (2002). An objective and method to rank the importance of the factors predisposing to landslides with the GIS methodology, application to an area of the Apennines (Valnerina; Perugia, Italy). *Engineering Geology, 63*(3-4), 277–289.

Ercanoglu, M., & Gokceoglu, C. (2004). Use of fuzzy relations to produce landslide susceptibility map of a landslide prone area (West Black Sea Region, Turkey). *Engineering Geology, 75*, 229–250.

Guzzetti, F., Carrara, A., Cardinali, M., & Reichenbach, P. (1999). Landslide Hazard Evaluation: A review of current techniques and their application in a multi-scale study, Central Italy. *Journal of Geomorphology, 31*, 181–216. London: Elsevier.

Intarawichian, N., & Dasananda, S. (2011). Frequency Ratio model based landslide susceptibility mapping in lower Mae Chaem watershed, Northern Thailand. *Environment and Earth Science, 64*, 2271–2285.

Ilanloo, M. (2011). A comparative study of fuzzy logic approach for landslide susceptibility mapping using GIS: An experience of Karaj dam basin in Iran. *Procedia - Social and Behavioral Sciences, 19*, 668–676.

Jibson, W. R., Edwin, L. H., & John, A. M. (2000). A method for producing digital probabilistic seismic landslide hazard maps. *Engineering Geology, 58*, 271–289.

Kanungo, D., Arrora, M., Sarkar, S., & Gupta, R. (2009). Landslide Susceptibility Zonation (LSZ) mapping -A review. *Journal of South Asia Disaster Studies, 2*, 81–105.

Karim, S., Jalileddin, S., & Ali, M. T. (2011). Zoning landslide by use of frequency ratio method (case study: Deylaman Region). *Middle-East Journal of Scientific Research, 9*(5), 578–583.

Lee, S., Ryu, J. H., Lee, M. J., & Won, J. S. (2003). Use of an artificial neural network for analysis of the susceptibility to landslides at Boun, Korea. *Environmental Geology, 44*, 820–833.

Lee, S., & Pradhan, B. (2006a). Landslide hazard assessment at Cameron Highland Malaysia using frequency ratio and logistic regression models. *Geophysical Research Abstracts, 8*, SRef ID: 1607-7962/gra/EGU06-A-03241.

Lee, S., & Sambath, T. (2006). Landslide susceptibility mapping in the DamreiRomel area, Cambodia using frequency ratio and logistic regression models. *Environmental Geology, 50*, 847–855.

Lee, S., & Talib, J. A. (2005). Probabilistic landslide susceptibility and factor effect analysis. *Environmental Geology, 47*, 982–990.

Lee, S., & Pradhan, B. (2006b). Probabilistic landslide risk mapping at Penang Island, Malaysia. *Journal of Earth System Science, 115*(6), 661–672.

Lee, S., & Pradhan, B. (2007). Landslide hazard mapping at Selangor, Malaysia using frequencyratio and logistic regression models. *Landslides, 4*(1), 33–41.

Mandal, S., & Maiti, R. (2011). Landslide susceptibility analysis of shivkhola watershed, darjeeling: A remote sensing & GIS based analytical hierarchy process (AHP). *Journal of Indian Society of Remote Sensing*. https://doi.org/10.10007/s12524-011-0160-9.

Mandal, S., & Maiti, R. (2013). Integrating the Analytical Hierarchy Process (AHP) and the Frequency Ratio (FR) model in landslide susceptibility mapping of Shiv-khola Watershed, Darjeeling Himalaya. *International Journal of Disaster Risk Science, 4*(4), 200–212. https://doi.org/10.1007/s13753-013-0021-y.

Mandal, B., & Mandal, S. (2016). Assessment of mountain slope instability in the lish river basin of eastern Darjeeling Himalaya using frequency ratio model (FRM) Model. *Earth Systems and Environment, 2*(121), 1–14. https://doi.org/10.1007/s40808-016-0169-8.

Mandal, B., & Mandal, S. (2017). Landslide susceptibility mapping using modified information value model in the lish river basin of Darjiling Himalaya. *Spatial Information Research, 2*(7), 1–14. https://doi.org/10.1007/s41324-017-0096-4.

Muthu, K., & Petrou, M. (2007). Landslide hazard mapping using an ExpertSystem and a GIS. *Transactions on Geoscience and Remote Sensing, 45*(2), 522–531.

Nithya, E. S., & Prasanna, R. P. (2010). An integrated approach with GIS and remote sensing technique for landslide zonation. *International Journal of Geomatics and Geosciences, 1*(1), 66–75.

Oliveira, S. C., Zêzere, J. L., Catalão, J., & Nico, G. (2015). The contribution of PSInSAR interferometry to landslide hazard in weak rocks dominated areas. *Landslides, 12*, 703–719.

Pandey, A., Dabral, P. P., Chowdhary, V. M., & Yadav, N. K. (2008). Landslide hazard zonation using remote sensing and GIS: A case study of Dikrong river basin, Arunachal Pradesh, India. *Environmental Geology, 54*, 1517–1529.

Pereira, S., Zezere, J., & Bateira, C. (2012). Assessing predictive capacity and conditional independence of landslide predisposing factors for shallow landslide susceptibility models. *Natural Hazards and Earth System Sciences, 12*, 979–988. https://doi.org/10.5194/nhess129792012.

Pistocchi, A., Luzi, L., & Napolitano, P. (2002). The use of predictive modelling techniques for optimal exploitation of spatial databases: A case study in landslide hazard mapping with expert system-like methods. *Environmental Geology, 41*, 765–775.

Poudyal, C. P., Chang, C., Oh, H. J., & Lee, S. (2010). Landslide susceptibility maps comparing frequency ratio and artificial neural networks: A case study from the Nepal Himalaya. *Environmental Earth Sciences, 61*, 1049–1064.

Pradhan, B., & Lee, S. (2010). Landslide susceptibility assessment and factor effect analysis: Backpropapagation artificial neural networks and their comparison with frequency ratio and bivariate logistic regression modelling. *Environmental Modelling & Software, 25*, 747–759.

Porghasemi, H. (2007). *Landslide hazard zoning statistical frequency ratio method in the basin Safarood.* M.Sc Thesis, TarbiatModarres University, Noor, p. 1386.

Pradhan, B. (2010). Remote sensing and GIS-based landslide hazard analysis and cross validation using multivariate logistic regression model on three test areas in Malaysia. *Advances in Space Research, 45*, 1244–1256.

Pradhan, B., & Lee, S. (2009). Delineation of landslide hazard areas using frequency ratio, logistic regression and artificial neural network model at Penang Island, Malaysia. *Environmental Earth Sciences, 60*, 1037–1054.

Rowbotham, D., & Dudycha, D. N. (1998). GIS Modelling of slope stability in Phewa Tal Watershed, Nepal. *Geomorphology, 26*, 151–170.

Sarkar, S., & Kanungo, D. P. (2004). An integrated approach for landslide susceptibility mapping using remote sensing and GIS. *Photogrammetric Engineering & Remote Sensing, 70*(5), 617–625.

Sarkar, S., Kanungo, D., Patra, A., & Kumar, P. (2006). *Disaster mitigation of debris flows, slope failures and landslides: GIS based landslide susceptibility mapping case study in Indian Himalaya* (pp. 617–624). Tokyo, Japan: Universal Academy Press.

Sharma, L., Patel, N., Ghosh, M., & Debnath, P. (2009). Geographical information system based landslide probabilistic model with trivariate approach - A case study in Sikkim Himalaya. Eighteenth United Nations Regional Cartographic Conference for Asia and the Pacific, UN, Bankok, Economic and Social Council.

Umar, Z., Pradhan, B., Ahmad, A., Jebur, M. N., & Tehrany, M. S. (2014). Earthquake induced landslide susceptibility mapping using an integrated ensemble frequency ratio and logistic regression models in West Sumatera Province, Indonesia. *Catena, 118*, 124–135.

Vijith, H., Rejith, P. G., & Madhu, G. (2009). Using Info Val Method and GIS techniques for the spatial modelling of landslides susceptibility in the Upper catchment of River Meenachil in Kerala. *Indian Society of Remote Sensing, 37*, 241–250.

Wang, H., & Sassa, K. (2005). Comparative evaluation of landslide susceptibility in Minamata area, Japan. *Environmental Geology, 47*, 956–966. https://doi.org/10.1007/s0025400512252.

Wang, Q., et al. (2015). Landslide susceptibility mapping based on selected optimal combination of landslide predisposing factors in a large catchment. *Sustainability, 7*, 16653–16669. https://doi.org/10.3390/su71215839.

Zezere, J. (2002). Landslide susceptibility assessment considering landslide typology: A case study in the area north of Lisbon (Portugal). *Natural Hazards and Earth System Sciences, 2*, 73–82. https://doi.org/10.5194/nhess-2-73-2002.

Zhou, C. H., Lee, C. F., Li, J., & Xu, Z. W. (2002). On the spatial relationship between landslide and causative factors on Lantau Island, Hong Kong. *Geomorphology, 43*, 197–207.

Chapter 4
Logistic Regression (LR) Model and Landslide Susceptibility: A RS and GIS-Based Approach

4.1 Introduction

Various quantitative and qualitative models and methods in connection to the landslide susceptibility assessment, prediction, and its spatial distribution were applied by researchers all over the world. To accomplish all the statistical models, geomorphic and hydrologic parameters were used scientifically. Logistic regression model is one of the most significant multi-variate statistical methods for landslide hazard zonation mapping for the mountain terrain. This method is dealt with the relationship between dependent variable and independent variables which are responsible for landslide occurrences. Geology, hydrology, hydrogeology, topography and morphology, climate and weathering play an important role in causing landslides (Soeters and van Westen 1996; Peart et al. 2005; Dominguez-Cuesta et al. 2007). Three important topographic factors like elevation, slope, and directionality of slope influence on landslide occurrences phenomena (Dai and Lee 2002; Nefeslioglu et al. 2008). Can et al. (2005), Lee and Sambath (2006), and Greco et al. (2007) suggested slope curvature as one of the most important topographic parameters of landslides. Slope instability takes place as a result of heavy or long duration rainfall, percolation of water into cracks (Crosta 1998; Dahal and Hasegawa 2008), existence of active faults (Ayalew et al. 2005), streams activities (Yaclin 2008), construction of roads and communication lines (Greco et al. 2007; Lee and Pradhan 2007), land use dynamics (Lee and Sambath 2006; Greco et al. 2007), and weak lithological composition of the region concerned (Yaclin 2008). The preparation of various geomorphic data layers, i.e. elevation, slope, aspect, curvature, topographic wetness index, and stream power index as well as their integration using a suitable algorithm have introduced a new dimension in the study of landslide susceptibility assessment and prediction. The introduction of quantitative approaches in geomorphological studies in the second half of twentieth century helped a lot to study earth surface processes and landform. Then, quantification and measurement of geomorphic attributes with the help of GIS have extended the nature and content of

S. Mandal, S. Mondal, *Statistical Approaches for Landslide Susceptibility Assessment and Prediction*, https://doi.org/10.1007/978-3-319-93897-4_4

scientific study of land surface. As the process of weathering and mass movement contribute a lot in changing character of the landform over the earth surface, the analysis of various factors responsible for the same processes have become much more significant to researchers. To assess and analyse all the contributing factors corresponding to weathering and mass movement processes, in the present day geographers, geologists, and geomorphologists applied geo-informatics as well as various models. Darjeeling Himalaya is one of the significant geomorphic units of the whole Himalayan mountain range where weathering and mass movement processes both contribute in drainage network development and landform evolution. The continuous growth and development of drainage network and slope evolution made the mountain slope of Darjeeling Himalaya more vulnerable to landslides. Not only that the expansion of human settlement and communication lines in Darjeeling Himalayan region have aggravated the problems of slope instability by putting enormous pressure over disintegrated and decomposed slope materials. Researchers have applied various statistical models to study geo-spatial distribution of slope instability over the space. Solaimani et al. (2012), Demir et al. (2014), Lee and Pradhan (2007), Bai et al. (2010), Wu et al. (2017) and Mondal and Mandal (2017) successfully applied logistic regression model for the preparation of landslide susceptibility zonation map of the concerned landslide-prone areas. In the present study, various landslide causative factors, i.e. elevation, slope aspect, slope angle, slope curvature, geology, soil, lineament density, distance to lineament, drainage density, distance to drainage, stream power index (SPI), topographic wetted index (TWI), rainfall, normalized differential vegetation index (NDVI), and land use and land cover (LULC) were considered because all these attributes have dominant role in slope instability of Darjeeling Himalaya. All these attributes were integrated on GIS environment using logistic regression (LR) model to assess the spatial distribution of landslide susceptibility.

4.2 Materials and Methods

To accomplish LR model and to prepare landslide susceptibility map of Darjeeling Himalaya, various thematic data layers such as slope, aspect, curvature, lithology, geomorphology, soil, distance to drainage, drainage density, distance to lineaments, lineament density, land use and land cover, NDVI, and rainfall were developed and their integration has been made on GIS platform. Major data sources are Landsat TM Image, Google earth image, topographical map (Survey of India), SRTM DEM (USGS Earth Explorer), geological map, geomorphological map, soil map (NATMO, National Atlas and Thematic Mapping Organisation), etc. (Table 4.1 and Fig. 4.1). To perform LR model, logistic regression co-efficient values were derived and on the basis of which landslide susceptibility map of Darjeeling Himalaya were made. To validate the result, training data set and validation data set were made based on the prepared landslide inventory map of Darjeeling Himalaya. The detailed description of the methods for the development of various data layers were elaborated in Chap. 2.

Table 4.1 Data layers, types, and sources of the study

Base map/image name	Source	Data layer	Data type
Topographical map and Google earth (2015–16)	Survey of India (SOI), GPS data and Internet	Landslide inventory map	ARC/INFO polygon and point coverage
Geological map	Geological Survey of India, Kolkata	Geological map	Categorical nominal data with ARC/INFO polygon coverage
Soil map	NBSS & LUP Regional Centre, Kolkata	Soil map	Categorical nominal data with ARC/INFO polygon coverage
SRTM DEM (30 m*30 m)	earthexplorer.usgs.gov	Elevation	Continuous raster data with ARC/INFO grid
		Aspect	Categorical nominal data with ARC/INFO grid
		Slope	Continuous raster data with ARC/INFO grid
		Curvature	Continuous raster data with ARC/INFO grid
		SPI	Continuous raster data with ARC/INFO grid
		TWI	Continuous raster data with ARC/INFO grid
Landsat TM image, Feb. 2009 (30 m*30 m)	glovis.usgs.gov	NDVI	Continuous raster data with ARC/INFO grid
Land use and land cover map	www.banglarbhumi.gov.in	LULC	Categorical nominal data with ARC/INFO polygon coverage
Drinking water prospects map	www.wbphed.gov.in	Lineament density	ARC/INFO line coverage
		Distance from lineament	ARC/INFO line coverage
Disaster mitigation map of Darjeeling district	http://darjeeling.gov.in/gismaps.html	Drainage density	ARC/INFO line coverage
		Distance to drainage	
Rainfall data (1950–2010)	http://www.worldclim.org	Rainfall map	Continuous raster data with ARC/INFO grid

4.2.1 Logistic Regression (LR) Model and Landslide Susceptibility Mapping

The logistic regression is one of the most popular methods which shows the relationship between a dependent variable and several independent variables (Pradhan and Lee 2010). It is introduced by McFadden (1974) for the measurement of probability of any hazard or disaster using specific equation and landslide controlling variables. The training data set of landslide area used in this study was polygon-based vector data layer, and this data set was converted into point data layer in Arc Map 10.1. Similarly, point data layers were incorporated within the non-landslide area of the

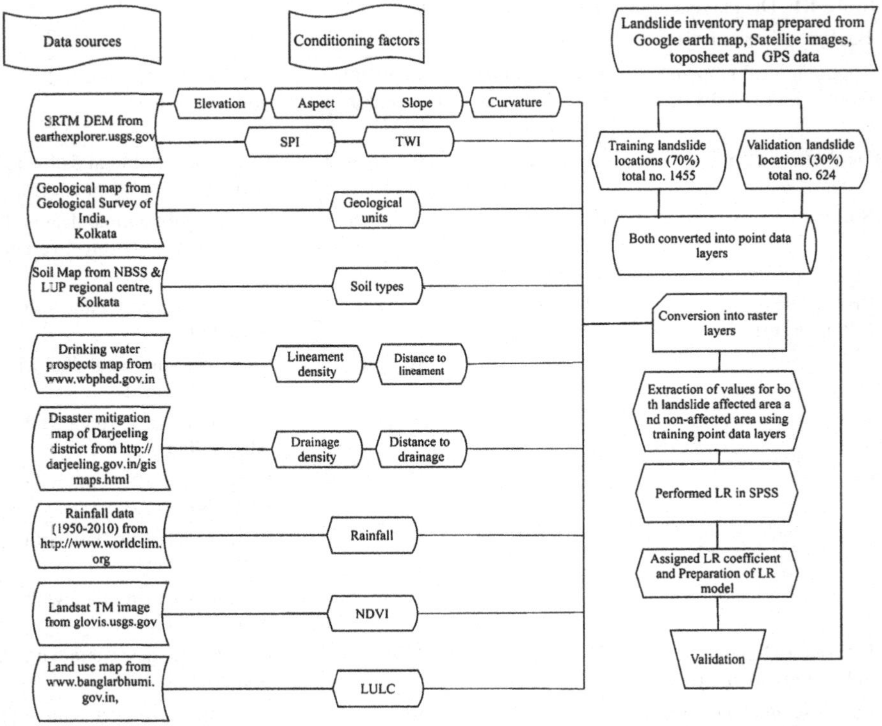

Fig. 4.1 Methodology of LR Model and development of landslide susceptibility map

basin which is known background points. Both the training landslide points and background points were used for the extraction of individual values of all the converted 30 m cell size raster independent layers. All the data values were imported in SPSS 17.0 and creating a new variable whose values were between 1 and 0. Extracted values by training points and background points were coded as 1 and 0 in new variable column, respectively. 1 means presence of landslide over the area and 0 means absence of landslide. Finally, the logistic regression was performed using SPSS 17.0.

The values of LR are directly related to the probability of landslide occurrence. The probability of landslide occurrence was calculated using the following equation (Eq. 4.1):

$$P = 1/\left(1 + e^{-z}\right) \quad \text{(Eq. 4.1)}$$

where p is the probability of landslide occurrence, e is the exponential, and z is the linear combination. The probability value ranges from 0 to 1 on an S-shaped curve. The linear combination "z" was shown in the following equation (Eq. 4.2).

$$z = b_0 + b_1 x_1 + b_2 x_2 + b_3 x_3 + b_4 x_4 + \ldots\ldots\ldots\ldots + b_n x_n \quad \text{(Eq. 4.2)}$$

where b_0 is the intercept or constant of the model, b_i ($i = 0, 1, 2, 3, 4,\ldots, n$) denotes the co-efficients of LR model and X_i = (i = 0, 1, 2, 3, 4,…, n) represents the influencing factors (Lee and Sambath 2006).

4.2.2 *Model Validation and Accuracy Assessment of Logistic Regression (LR) Model*

The assessment of the effectiveness of produced landslide susceptibility map is the important part of landslide susceptibility mapping. According to Can et al. (2005), two decision rules were considered for the assessment of performance of landslide susceptibility map. These are:

- Decision rule 1: Most of the observed landslides should have to be found in high susceptibility class.
- Decision rule 2: High susceptibility classes should have to cover small area as possible on the prepared susceptibility map.

Beside this, receiver operating characteristics (ROC) (Falaschi et al. 2009; Das et al. 2012; Park et al. 2013; Pourghasemi et al. 2013) was used for the accuracy assessment of the landslide susceptibility map. The ROC curve is a graphical representation of the false positive rate (FPR) on the X axis (Eq. 4.3) and the true positive rate (TPR) on the Y axis (Eq. 4.4).

$$X = \text{FPR} = 1 - \frac{\text{TN}}{\text{TN} + \text{FP}} \qquad \text{(Eq. 4.3)}$$

$$Y = \text{TPR} = 1 - \frac{\text{TP}}{\text{TP} + \text{FN}} \qquad \text{(Eq. 4.4)}$$

where TN is the true negative, FP is the false positive, TP is the true positive, and FN is the false negative. It has the ability to predict an event. AUC value of ROC curve ranges from 0.5 to 1.0. AUC 1.0 means perfect prediction. The values of ROC curve can be classified into five classes on the basis of quantitative–qualitative relationship between AUC value and prediction accuracy, i.e. 0.9–1.0 = excellent, 0.8–0.9 = very good, 0.7–0.8 = good, 0.6–0.7 = average, and 0.5–0.6 = poor (Yesilnacar 2005). In an ideal landslide susceptibility map, the frequency ratio value is decreasing from very high to very low susceptibility zones (Pourghasemi et al. 2012).

4.3 Result and Discussion

4.3.1 *Model Validation*

The LR model was assessed with the variables of 0 and 1 which were determined based on one or several independent variables (Menard 2001). "0" indicates absence whereas "1" means presence with respect to an event. To estimate LR regression co-efficient value, the following equation (Eq. 4.5) was developed and performed.

$$
\begin{aligned}
z = -21.860 + (\text{"}SPI\text{"} \times 0.332) + (\text{"Aspect"} \times -0.003) \\
+ (\text{"Curvature"} \times -0.018) + (\text{"}TWI\text{"} \times -0.0001) \\
+ (\text{"Slope"} \times 0.090) + (\text{"Drainage density"} \times -0.071) \\
+ (\text{"Elevation"} \times -0.0003) + (\text{"Lineament density"} \times 0.223) \\
+ (\text{"Rainfall"} \times 0.001) + (\text{"Distance to Drainage"} \times -0.080) \\
+ (\text{"Distance to Lineament"} \times 0.00003) \\
+ (\text{"NDVI"} \times -8.670) + \text{"LULC"} + \text{"Geology"} + \text{"Soil"}
\end{aligned}
\quad \text{(Eq. 4.5)}
$$

Using the above equation, the landslide probability value was derived incorporating the probability of landslide occurrences formula on Arc Map 10.1 and finally the landslide probability map was obtained, and it was classified into five, i.e. very low, low, moderate, high, and very high landslide susceptibility. Altitude, slope aspect, curvature, drainage density, distance to drainage, topographic wetness index (TWI), and NDVI showed negative regression co-efficient value, and it is assumed that all these parameters played less significant role in landslide probability. Slope, most of the geological unit of Darjeeling Himalaya (except alluvium), all the soil groups, distance to lineaments, lineament density, stream power index, and rainfall are showing positive logistic regression co-efficient values (Table 4.2) and indicating all these factors as most significant in landslide susceptibility studies.

The study revealed that Darjeeling Himalaya is dominated by moderate landslide susceptibility (25.48%) and which is followed by low (22.24%), high (20.54%), very low (18.95%), and high (12.79%) landslide susceptibility (Fig. 4.2). It was found that 37.35% landslide affected areas are being found in very high landslide susceptibility zones, 26.54% landslide affected area in high landslide susceptibility zone, and 17.85% landslide affected area in moderate landslide susceptibility zone (Table 4.3). There is a positive relationship between landslide susceptibility and frequency ratio values. Higher the landslide susceptibility, greater is frequency ratio or landslide probability in Darjeeling Himalaya. The frequency ratio in each landslide susceptibility class depicts the reasonable accuracy of LR model and where the real validity of the model exists.

4.3.2 *Logistic Regression (LR) Model Validation by ROC Curve and Frequency Ratio*

The area under curve (AUC) of the ROC curve was 0.715 or 71.5% (Fig. 4.3) which states that the logistic regression (LR) method used in the study showed a reasonable accuracy in predicting the landslide susceptibility of Darjeeling Himalaya. The frequency ratio plot (Fig. 4.4) showed that the frequency ratio was gradually increased from the moderate to the very high susceptibility zone in which also indicated a reasonable accuracy of the prepared landslide susceptibility map of Darjeeling Himalaya.

Table 4.2 Logistic regression co-efficients and test statistics of the landslide conditioning factors

Conditioning factor	Class	Logistic co-efficient [a] (*B*)	Standard error of estimate (S.E.)	Wald Chi-square value	Significance probability (Sig.)
Altitude (m)	130–470	−0.0003	0.000	261.858	0.000
	470–770				
	770–1100				
	1100–1300				
	1300–1600				
	1600–1900				
	1900–2200				
	2200–2600				
	2600–3000				
	3000–3600				
Aspect	Flat (−1)	−0.003	0.000	1791.608	0.000
	North (0–22.5)				
	Northeast (22.5–67.5)				
	East (67.5–112.5)				
	Southeast (112.5–157.5)				
	South (157.5–202.5)				
	Southwest (202.5–247.5)				
	West (247.5–292.5)				
	Northwest (292.5–337.5)				
	North (337.5–360)				
Slope (Degree)	0–4.339	0.090	0.001	6108.478	0.000
	4.339–8.677				
	8.677–12.515				
	12.515–15.853				
	15.853–19.024				
	19.024–22.027				
	22.027–25.031				
	25.031–28.202				
	28.202–32.373				
	32.373–42.553				
Curvature	−0.644 to −0.238	−0.018	0.093	0.040	0.842
	−0.238 to −0.155				
	−0.155 to −0.095				
	−0.095 to −0.040				
	−0.040 to 0.015				
	0.015–0.064				
	0.064–0.120				
	0.120–0.185				
	0.185–0.267				
	0.267–0.756				

(continued)

Table 4.2 (continued)

Conditioning factor	Class	Logistic co-efficient [a] (*B*)	Standard error of estimate (S.E.)	Wald Chi-square value	Significance probability (Sig.)
Geology	Alluvium	−1.706	0.123	193.601	0.000
	Baxa Series-Slates, Schists, Dolomite, Quartzites	0.858	0.161	28.513	0.000
	Daling Series-Slates, Schists, Quartzites	0.027	0.034	0.634	0.426
	Damuda (Gondwana)	1.012	0.055	333.352	0.000
	Darjeeling Gneiss	0.365	0.039	87.296	0.000
	Nahan Group (Tertiary) Limestone	–	–	–	–
Soil	Tista River	19.321	3498.200	0.000	0.996
	W001	21.679	3498.200	0.000	0.995
	W002	20.825	3498.200	0.000	0.995
	W003	21.508	3498.200	0.000	0.995
	W004	22.046	3498.200	0.000	0.995
	W006	17.532	3498.200	0.000	0.996
	W007	19.555	3498.200	0.000	0.996
	W008	19.197	3498.200	0.000	0.996
	W009	–	–	–	–
Drainage density (km/Sq. km)	0–0.694	−0.071	0.011	42.760	0.000
	0.694–1.210				
	1.210–1.620				
	1.620–2.010				
	2.010–2.370				
	2.370–2.730				
	2.730–3.140				
	3.140–3.680				
	3.680–4.450				
	4.450–6.560				
Lineament density (km/Sq. km)	0–0.056	0.223	0.041	29.985	0.000
	0.056–0.168				
	0.168–0.287				
	0.287–0.399				
	0.399–0.505				
	0.505–0.610				
	0.610–0.716				
	0.716–0.835				
	0.835–0.966				
	0.966–1.590				

(continued)

Table 4.2 (continued)

Conditioning factor	Class	Logistic co-efficient [a] (*B*)	Standard error of estimate (S.E.)	Wald Chi-square value	Significance probability (Sig.)
Distance from lineament (m)	0–20	0.00003	0.000	14.644	0.000
	20–50				
	50–100				
	100–400				
	400–1000				
	1000–2000				
	2000–3500				
	3500–5000				
	5000–6500				
	6500–8500				
Distance to drainage (m)	0–10	−0.080	0.006	176.275	0.000
	10–30				
	30–70				
	70–150				
	150–300				
	300–700				
	700–1200				
	1200–1700				
	1700–2300				
SPI	−5.939 to −3.140	0.332	0.012	707.832	0.000
	−3.140 to 2.496				
	−2.496 to −2.013				
	−2.013 to −1.595				
	−1.595 to −1.209				
	−1.209 to 0.791				
	−0.791 to −0.373				
	−0.373 to 0.014				
	0.014 to 0.400				
	0.400 to 2.266				
TWI	−6907.755	−0.0001	0.000	0.864	0.353
	−6907.755 to −2849.717				
	−2849.717 to −1226.502				
	−1226.502 to −520.757				
	−520.757 to −167.884				
	−167.884 to −26.735				
	−26.735 to 43.840				
	43.840–255.564				
	255.564–714.298				
	714.298–2090.502				

(continued)

Table 4.2 (continued)

Conditioning factor	Class	Logistic co-efficient [a] (*B*)	Standard error of estimate (S.E.)	Wald Chi-square value	Significance probability (Sig.)
Avg. annual rainfall (mm)	1076–1478	0.001	0.000	442.102	0.000
	1478–1798				
	1798–2097				
	2097–2355				
	2355–2551				
	2551–2705				
	2705–2860				
	2860–3056				
	3056–3293				
	3293–3705				
NDVI	−0.220 to 0.037	−8.670	0.073	14090.559	0.000
	0.037–0.110				
	0.110–0.166				
	0.166–0.214				
	0.214–0.263				
	0.263–0.315				
	0.315–0.368				
	0.368–0.427				
	0.427–0.493				
	0.493–0.667				
LULC	Social Forestry	−1.043	0.025	1794.264	0.000
	Agricultural Single Crop	−1.163	0.024	2281.963	0.000
	Forest	–	–	–	–
	Rural Settlement	−2.742	0.074	1391.545	0.000
	Urban Settlement	−3.774	0.313	145.854	0.000
	Agricultural Plantation (Tea)	−0.732	0.027	745.788	0.000
	Wasteland Dry/ Barren Land	1.156	0.337	11.761	0.001
	Public Utility and Facility	−18.707	2857.153	0.000	0.995
	Wastelands with Scrub	1.215	0.055	492.521	0.000

[a]Constant: −21.860

4.3.3 Validation of the Model Based on Independent Variables

If all the variables are included in the LR model (altitude, slope aspect, slope angle, slope curvature, geology, soil, LULC, NDVI, drainage density, lineament density, distance from lineament, distance to drainage, SPI, TWI, and rainfall), −2 Log likelihood value is the smallest (116024.695) whereas Cox & Snell R Square and

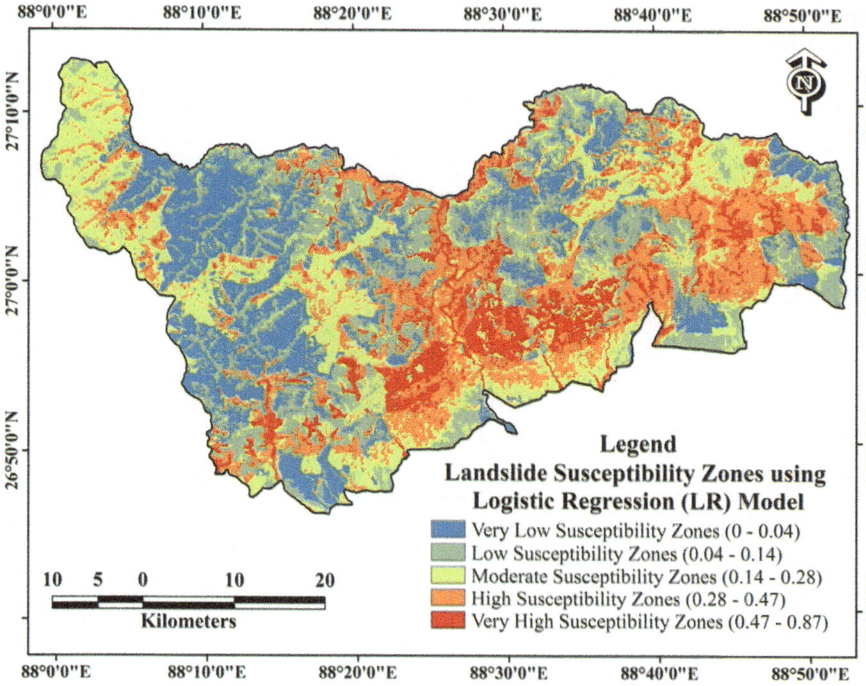

Fig. 4.2 Landslide susceptibility map of Darjeeling Himalaya

Table 4.3 Statistics of different susceptibility classes of the prepared landslide susceptibility map using LR model

Landslide susceptibility zones	Total pixels	% of total pixels	Landslide pixels	% of landslide pixels	Frequency ratio value
Very low	488,555	18.95	293	6.01	0.32
Low	573,375	22.24	598	12.25	0.55
Moderate	656,907	25.48	871	17.85	0.70
High	529,547	20.54	1296	26.54	1.29
Very high	329,742	12.79	1823	37.35	2.92

Nagelkerke R Square values are largest, that is, 0.343 and 0.485, respectively, which means that the factors responsible for landslide occurrences are statistically significant and LR model is considered as successful (Table 4.4).

4.4 Conclusion

The landslide susceptibility map of Darjeeling Himalaya was classified into five groups, i.e. very low, low, moderate, high, and very high landslide susceptibility. The study revealed that Darjeeling Himalaya is dominated by moderate to high

Fig. 4.3 ROC curve and accuracy result

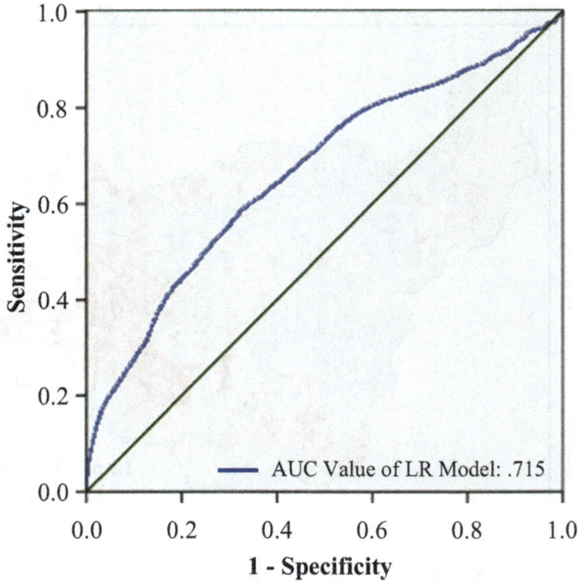

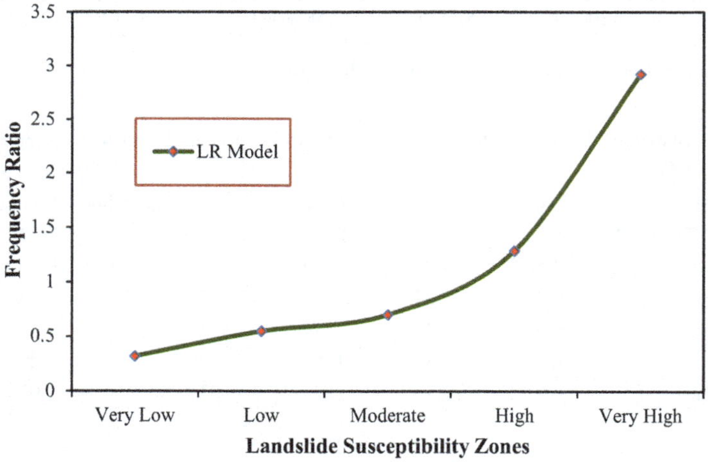

Fig. 4.4 Frequency ratio plot of different landslide susceptibility zones using LR Model

landslide susceptibility. Basically, rugged terrain, drainage concentration, weak lithology, and high intensity of rainfall contributed a suitable environment to make the slope more unstable in Darjeeling Himalaya. To perform LR model, 15 landslide conditioning factors such as elevation, slope aspect, slope angle, slope curvature, geology, soil, lineament density, distance to lineament, drainage density, distance to drainage, stream power index (SPI), topographic wetted index (TWI),

Table 4.4 Comparison of the performance of the LR models based on the number of independent variables

Independent variables	−2 Log likelihood	Cox & Snell R square	Nagelkerke R square
All 15 variables	116024.695	0.343	0.485
Except altitude	116288.045	0.342	0.483
Except aspect	117861.441	0.335	0.473
Except slope	122702.869	0.312	0.441
Except curvature	116024.735	0.342	0.484
Except geology	116937.301	0.339	0.479
Except soil	122430.813	0.313	0.443
Except land use/land cover	123125.744	0.310	0.438
Except NDVI	133883.912	0.256	0.362
Except drainage density	116067.616	0.342	0.484
Except lineament density	116054.579	0.342	0.484
Except distance to lineament	116039.299	0.342	0.484
Except distance to drainage	116200.385	0.342	0.484
Except SPI	116747.555	0.340	0.480
Except TWI	116025.531	0.342	0.484
Except rainfall	116467.038	0.341	0.482

rainfall, normalized differential vegetation index (NDVI) and land use and land cover (LULC) were made with the help of RS and GIS. To assess the relationship between causative factors and landslides, LR co-efficient values were derived. Positive LR values in favour of soil, distance to lineament, lineament density, SPI, TWI, and rainfall showed a strong influence on slope instability. Land use and land cover type of barren land (LR-1.156) and wasteland with scrubs (LR-1.215) have great impact on landslides in Darjeeling Himalaya. The area under ROC curve and frequency ratio plot depicted a reasonable accuracy of LR model. The prepared landslide susceptibility zonation map of Darjeeling Himalaya could provide a valid document for regional planning and development. The rapid rate of urbanization in Darjeeling Himalaya and associated infrastructural developmental activities, i.e. expansion of settlement, construction of roads, establishment of pipe lines for water, and expansion and development of tea garden area posed enormous pressure on rugged terrain and introduced suitable environment for the movement of slope materials downward. The movement of slope materials and lowering of the relief invite geomorphic equilibrium over the earth surface. Human intervention played a significant role in restricting geomorphic equilibrium by clearing the landslide materials from the foothills environment, constructing various retaining structures and modifying the slope surface. In Darjeeling Himalaya, most landslide affected areas of Darjeeling town, Kurseon town, Kalimpong town, Mirik town, and its surrounding regions are characterized by high landslide susceptibility and where human intervention is at greater scale and where geomorphic equilibrium requires longer temporal scale.

References

Ayalew, L., Yamagishi, H., Marui, H., & Kanno, T. (2005). Landslides in Sado island of Japan: part II, GIS-based susceptibility mapping with comparisons of results from two methods and verifications. *Engineering Geology, 81*, 432–445.

Bai, S., Lü, G., Wang, J., Zhou, P., & Ding, L. (2010). GIS-based rare events logistic regression for landslide-susceptibility mapping of Lianyungang, China. *Environmental Earth Sciences, 62*(1), 139–149.

Can, T., Nefeslioglu, H. A., Gokceoglu, C., Sonmez, H., & Duman, T. Y. (2005). Susceptibility assessments of shallow earth flows triggered by heavy rainfall at three catchments by logistic regression analyses. *Geomorphology, 72*, 250–271.

Crosta, G. (1998). Regionalization of rainfall thresholds: an aid to landslide hazard evaluation. *Environmental Geology, 35*, 131–145.

Dahal, R. K., & Hasegawa, S. H. (2008). Representative rainfall thresholds for landslides in the Nepal Himalaya. *Geomorphology, 100*, 429–443.

Dai, F. C., & Lee, C. F. (2002). Landslide characteristics and slope instability modelling using GIS, Lantau Island, Hong Kong. *Geomorphology, 42*, 213–228.

Das et al. (2012). Landslide susceptibility along road corridors in the Indian Himalaya using Bayesian logistic regression models. *Geomorphology, 179*, 116–125.

Demir, G., Aytekin, M., & Akgun, A. (2014). Landslide susceptibility mapping by frequency ratio and logistic regression methods: An example from Niksar–Resadiye (Tokat, Turkey). *Arabian Journal of Geosciences*. https://doi.org/10.1007/s12517-014-1332-z.

Dominguez-Cuesta, M., Jimenez-Sonchez, M., & Berrezueta, E. (2007). Landslide in the central coalfield (Cantabarian Mountains, NW Spain): Geomorphological feature conditioning factors and meteorological implication in susceptibility assessment. *Geomorphology, 89*, 1–12.

Falaschi et al. (2009). Logistic regression versus artificial neural networks: Landslide susceptibility evaluation in a sample area of the Serchio River valley, Italy. *Natural Hazards, 50*, 551–569.

Greco, R., Sorriso, V., & Catalano, E. (2007). Logistic regression analysis in the evaluation of mass movement's susceptibility case study: Calabria, Italy. *Engineering Geology, 89*, 47–66.

Lee, S., & Pradhan, B. (2007). Landslide hazard mapping at Selangor Malaysia using frequency ratio and logistic regression models. *Landslides, 4*, 33–41.

Lee, S., & Sambath, T. (2006). Landslide susceptibility mapping in the DamreiRomel area, Cambodia using frequency ratio and logistic regression models. *Environmental Geology, 50*, 847–855.

McFadden, D. (1974). Conditional logit analysis of qualitative choice analysis. In P. Zarembka (Ed.), *Frontiers in econometrics* (pp. 105–142). New York: Academic Press.

Menard, S. W. (Ed.). (2001). *Applied logistic regression analysis* (2nd ed., 111pp ed.). Thousand Oaks, CA: Sage.

Mondal, S., & Mandal, S. (2017). RS & GIS-based landslide susceptibility mapping of the Balason River basin, Darjeeling Himalaya, using logistic regression (LR) model, *Georisk: Assessment and Management of Risk for Engineered Systems and Geohazards, 12*(1), https://doi.org/10.1080/17499518.2017.1347949

Nefeslioglu, H. A., Gokceoglu, C., & Sonmez, H. (2008). An assessment on the use of logistic regression and artificial neural networks with different sampling strategies for the preparation of landslide susceptibility maps. *Engineering Geology, 97*, 171–191.

Park et al. (2013). Landslide susceptibility mapping using frequency ratio, analytical hierarchy process, logistic regression and artificial neural network methods at the Inje Area, Korea. *Environment and Earth Science, 68*(5), 1443–1464.

Peart, M. R., Ng, K. Y., & Zhang, D. D. (2005). Landslides and sediment delivery to a drainage system: some observations from Hong Kong. *Journal of Asian Earth Sciences, 25*, 821–836.

Pourghasemi, H. R., Pradhan, B., & Gokceoglu, C. (2012). Application of fuzzy logic and analytical hierarchy process (AHP) to landslide susceptibility mapping at Haraz watershed, Iran. *Natural Hazards, 63*, 965–996.

Pourghasemi, H. R., et al. (2013). Landslide susceptibility mapping using certainty factor, index of entrophy and logistic regression models in GIS and their comparison at Mugling-Narayanghat road section in Nepal Himalaya. *Natural Hazards, 65*(1), 135–165.

Pradhan, B., & Lee, S. (2010). Regional landslide susceptibility analysis using back propagation neural network model at Cameron Highland, Malaysia. *Landslides, 7*, 13–30.

Soeters, R., & van Westen, C. J. (1996). Slope instability recognition, analysis and zonation. In: Turner, A. K., & Schuster, R. L. (Eds.), *Landslides investigation and mitigation* (pp. 129–177). Transportation Research Board, National Research Council, Special Report 247. Washington, DC: National Academy Press.

Solaimani, K., Mousavi, S. Z., & Kavian, A. (2012). Landslide susceptibility mapping based on frequency ratio and logistic regression models. *Arabian Journal of Geosciences, 6*, 2557–2569. https://doi.org/10.1007/s12517-012-0526-5.

Wu, Z., Wu, Y., Yang, Y., Chen, F., Zhang, N., Ke, Y., & Li, W. (2017). A comparative study on the landslide susceptibility mapping using logistic regression and statistical index models. *Arabian Journal of Geosciences, 10*, 187. https://doi.org/10.1007/s12517-017-2961-9.

Yaclin, A. (2008). GIS-based landslide susceptibility mapping using analytical hierarchy process and bivariate statistics in Ardesen (Turkey): Comparisons of results and confirmations. *Catena, 72*, 1–12.

Yesilnacar, E. K. (2005) *The application of computational intelligence to landslide susceptibility mapping in Turkey*. Ph.D Thesis, Department of Geomatics, the University of Melbourne, p. 423.

Chapter 5
Artificial Neural Network (ANN) Model and Landslide Susceptibility

5.1 Introduction

Darjeeling Himalaya is one of the most landslide-prone areas of India as it causes great damages to human lives and properties frequently in every rainy period. To manage or mitigate the destructive impact of landslides, geologists, geographers, planners, and policy-makers started to introduce various methods and techniques in the late second half of twentieth century. Identification of landslide affected areas and its mapping became an important approach to landslide studies in India. Firstly, landslide hazard evaluation factor rating approach was applied for landslide hazard zonation mapping. But, this approach did not consider the accuracy assessment of the prepared landslide zonation map. Later on weighted overlay model, frequency ratio model, analytical hierarchy process, information value model, slope stability model, rainfall-intensity model, logistic regression model, and bi-variate and multivariate statistical models became much more significant in landslide susceptibility assessment and prediction in Darjeeling Himalaya of India. In Darjeeling Himalaya, geology, rainfall, drainage network development, and human intervention are the major landslide conditioning factors. Besides all these, slope, elevation, relief, slope curvature, land use and land cover also played dominating role in landslides. Starkel (1972) and Froehlich et al. (1990) investigated the impact of rainfall on landslide in Darjeeling. Several researchers such as Wilson and Wieczorek (1995), Wieczorek (1987, 1996), Terlien (1997, 1998), Crosta (1998), Crozier (1999), Glade et al. (2000), Aleotti (2004), Guzzetti et al. (2004), Guzzetti et al. (2007), Hong et al. (2005), and Zezere et al. (2005) established the implication of rainfall-intensity model for slope failure. Montgomery and Dietrich (1994) introduced a physically based model for the topographic control on shallow landsliding in terms of geomorphic threshold. Guzzetti et al. (2007) reviewed rainfall thresholds for the initiation of landslides worldwide and proposed new empirical thresholds based on the statistical analysis. The geomorphic thresholds (threshold slope, threshold height, and threshold rainfall) are the dominant contributing factors of slope instability but the

S. Mandal, S. Mondal, *Statistical Approaches for Landslide Susceptibility Assessment and Prediction*, https://doi.org/10.1007/978-3-319-93897-4_5

preparation and identification of susceptible landslide locations applying various statistical approaches and using RS and GIS were being accepted worldwide as a significant tool in landslide studies. In the present work, artificial neural network (ANN) model has been applied to prepare landslide susceptibility zones of Darjeeling Himalaya using geospatial tools. Artificial neural network model was successfully used by Zhou (1999), Pistocchi et al. (2002), Lee et al. (2003a, b) and Lee et al. (2006). Pradhan and Lee (2010a), Pradhan and Lee (2010b), and Pradhan and Lee (2010c) introduced frequency ratio model, logistic regression model, and artificial neural network model for landslide susceptibility assessment.

The region represents a complicated geological and tectonic milieu with rocks of varying ages and metamorphisms. To the south of Darjeeling, coarse to a very coarse-grained clastics (conglomerate-sandstone-siltstone) of the Siwalik Group of Tertiary age are exposed. The tectonic units in the area occur in the reverse order of stratigraphic superposition. In the study area, the Daling group is comprised of low-grade metamorphic rocks, while the Darjeeling Group consists primarily of gneisses. The presence of lineaments, hot springs, lakes, waterfalls, abrupt changes in river gradient, and erraticity in terrace distribution suggested that the area has been undergoing differential uplift along a number of regional and local sub-vertical faults (Banerji et al. 1980). The fragile lithological composition along with lineaments have provided suitable condition for continuous branching of the drainage network which helped in steepening the slope gradient as well as in making the relief more rugged in nature. It was also observed that the concentration of drainage over the space played an important role in landslide processes in Darjeeling Himalaya. The present study analysed landslide causative factors of elevation, slope aspect, slope angle, slope curvature, geology, soil, lineament density, distance to lineament, drainage density, distance to drainage, stream power index (SPI), topographic wetted index (TWI), rainfall, normalized differential vegetation index (NDVI) and land use and land cover (LULC) and finally an integration has been made on GIS environment using artificial neural network model (ANN).

The aim of this research is to identify landslide susceptibility zones with the Artificial Neural Networks (ANNs), which is associated with black box models to be implemented, similar to several other statistical approaches (Carrara et al. 1991, 1995). At present, ANN is used for various scientific and engineering applications (Emami et al. 1998; Caparrini et al. 1996) which begins from the database containing the variables that describe that particular process. ANN is applied to assess indirect determination of the triggering parameters (Mayoraz et al. 1996) as well as to assess landslide susceptibility mapping, with physical terrain factors (Lee et al. 2001). In recent studies, many mathematical techniques such as decision tree algorithm (Hong et al. 2015; Tsangaratos and Ilia 2016; Wu et al. 2014); fuzzy logic algorithm (Ilanloo 2011; Zhu et al. 2014; Bui et al. 2012); neuro-fuzzy algorithm (Sezer et al. 2011; Bui et al. 2012) have been introduced for evaluating landslide susceptible areas. Pradhan and Lee (2010a) evaluated three models for landslide susceptibility analysis using frequency ratio, logistic regression, and artificial neural network model.

5.2 Materials and Methods

5.2.1 *Artificial Neural Networks*

Artificial Neural Network (ANN) is considered as a machine which is designed in such a way that the human mind works to a specific task (Haykin 1999). It has the capability of learning and generalizing from the experience. The structure of ANN model: first one, the neurons which process the stimulus; second one, each neuron that has the ability to receive and transmit a signal is called the stimulus. Next one, each signal—stimulus received or transmitted from one neuron to another in the neural network associated with a weight (synaptic weight) which indicates the connective strength of the respective neurons. The weight value is positively related to the contribution of the node (Fausett 1994). Even if the forms of relationship of the set of input factors are unknown or their physical explanation is hard to be described, yet ANN model has the ability to predict based on the input data. This characteristic of ANN model makes more superior than the empirical and statistical methods which need the prior knowledge of the input data and the nature of the relationship that is linear, non-linear, etc.

5.2.2 *The Learning Process Method*

Through the iterative learning process, the neural networks receive stimuli (information and knowledge), as people do, and knowledge is stored in the network connections (Haykin 1999). The ANN models follow the parallel distributed processing which tries to combine the thinking of the human brain with the abstract mathematical thinking (McClelland et al. 1986). The learning method is classified into three types. Firstly, the supervised learning in which learning is accomplished by presenting a set of training patterns each with an associated target output vector. Secondly, the unsupervised learning where learning is accomplished by grouping a similar set of input patterns together without the use of training data to specify what a typical member of each group looks like or to which group each pattern belongs to (Fausett 1994). Last one, the reinforcement learning is learning by interacting with an environment. The most typical ANN setting is the one that enables supervised training. During the training phase, the hidden and output layer neurons process their inputs by multiplying each input by a corresponding weight, summing the product, and then processing the sum using a non-linear transfer function to produce a result. An ANN learns by adjusting the weights between the neurons in response to the errors between the actual output values and the target output values. At the end of this training phase, the neural network provides a model that should be able to predict a target value from a given input value.

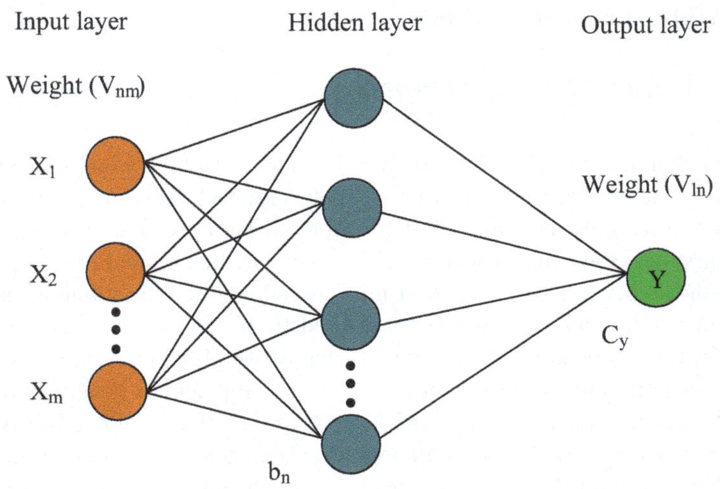

Fig. 5.1 The feed-forward multi-layer artificial neural network

5.2.3 The Basic Features of Artificial Neural Network (ANN) Model

According to Fausett (1994) and Benardos and Benardos (2005), each neural network has three basic features that is network architecture, training or learning algorithm, and activation function. Network architecture is the way in which the neurons are connected to each other. It is critical process and requires possible knowledge of the problem application (Benardos and Kaliampakos 2004) while training or learning algorithm is the method which establishes the values of the weights on the connections. Last one is the type of activation function used. Neural networks can be single layer or multi-layer which is examples of feed-forward networks in which the signals flow from the input units to the output units in a forward direction. In multi-layer perceptron (MLP) networks, there is always an input layer, a hidden layer, and an output layer which is used in this study (Fig. 5.1). The first layer of the network or input layer contains a node for each of the input variables. The second layer, the hidden layer, consists of nodes that allow complexities to develop amongst input nodes. The last layer of the network, or output layer, contains *t* nodes, one for each output type.

5.2.4 The Back-Propagation Algorithm

The training of an ANN by back-propagation learning algorithm has three stages that is the feed forward of the input training pattern, the calculation and back-propagation of the associated error, and the adjustment of the weights. In the first stage, each input unit receives an input signal and transmits this signal to each of the hidden units. Each hidden unit then calculates its activation and transmits its signal to each output unit (Eq. 5.1). Each output unit computes its activation to form the response of the net for the given input pattern (Eq. 5.2). Each neuron in the network may employ a non-linear activation function at the output end, producing smooth signals to other neurons. One of most commonly used activation functions is the binary sigmoid transfer function which has range of (0, 1) and is defined by as Hagan et al. (1996) (Eq. 5.3). Each output unit compares its activation with its target value to determine the associated error for that pattern with that unit (Eq. 5.4). Based on this error, a δ factor, is computed which is used to distribute the error at output unit back to all units in the hidden and input layer (Eqs. 5.5 and 5.6). After the entire δ factors have been calculated, the weights for all layers are adjusted simultaneously, according to the generalized Least-Square-Mean rule (Hagan et al. 1996) (Eq. 5.7).

$$\text{net}_j^{(l)}(t) = \sum_{i=0}^{p}\left(y_i^{(l-1)}(t)\,w_{ji}^{l}(t)\right), \text{the net input of } j\text{th neuron of layer } l \text{ and } t \text{ iteration} \quad \text{(Eq. 5.1)}$$

$$y_j^{l}(t) = f\left(\text{net}_j^{(l)}(t)\right) \quad \text{(Eq. 5.2)}$$

$$f(\text{net}) = \frac{1}{1+e^{(-\text{net})}} \quad \text{(Eq. 5.3)}$$

$$e_j(t) = c_j(t) - a_j(t) \quad \text{(Eq. 5.4)}$$

$$\delta_j^{(l)}(t) = e_j^{(l)}(t)\,a_j(t)\left[1 - a_j x(t)\right], \delta \text{ factor for neuron } j\text{th in the output layer } i\text{th} \quad \text{(Eq. 5.5)}$$

$$\delta_j^{(l)}(t) = y_j^{(l)}(t)\left[1 - y_j(t)\right]\Sigma\,\delta_j^{(l)}(t)\,w_{kj}^{(l+1)}(t), \delta \text{ factor for neuron } j\text{th in hidden layer } i\text{th}\ldots \quad \text{(Eq. 5.6)}$$

$$w_{ji}^{(l)}(t+1) = w_{ji}^{(l)}(t) + \alpha\left[w_{ji}^{(l)}(t) - w_{ji}^{(l)}(t-1)\right] + n\delta_j^{(l)}(t)\,y_j^{(l-1)}(t) \quad \text{(Eq. 5.7)}$$

where η is the learning rate, and α is the momentum rate for speeding up learning without running into the risk of oscillation.

5.2.5 *The Process of Landslide Susceptibility Analysis through Artificial Neural Network (ANN) Model in MATLAB Environment*

The probabilities of occurrence of landslides were calculated based on the various input attributes and their cumulative influence and knowledge-based classification. Before running the artificial neural network program, the training site should be selected. So, the landslide prone (occurrence) area and the landslide-not-prone area were selected as training sites. There is no precise mathematical law to establish the required minimum size of landslide prone data sets. In this study, about 70% of the data set is usually considered as sufficient for training the network and the rest of it is normally reserved for testing the model. Pixels from each of the two classes were randomly selected as training pixels denoting areas where landslide not occurred or occurred. First, areas where the landslide was not occurred were classified as "areas not prone to landslide" and areas where landslide was known to exist were assigned to an "areas prone to landslide" training set. Training sites were selected based on landslide location as prone training site and with a varying slope values as non-prone training site and then the MLP trained back-propagation algorithm was computed. Hidden layers were selected two times of input attribute layers. So obviously, the output will have both "existing" and "non-existing" landslide areas. Some of the input attribute layers are continuous and others categorical in nature. Therefore, these data were converted to raster grid in order to apply the ANN model. The model had 15 neurons in the input layer, with 30 hidden neurons. The 15 neurons in the input layer correspond to the landslide-related factors (elevation, slope, aspect, curvature, lineament, distance to lineament, drainage density, distance to drainage, NDVI, land use and land cover, geology, soil, rainfall, SPI, and TWI) that had been identified as causative factors. Three-layered feed-forward network was implemented using the MATLAB software package. The number of hidden layers and the number of nodes in a hidden layer required for a particular classification problem are not easy to deduce. In this study, an input layer (15 input neurons) × one hidden layer (30 hidden neurons) × one output layer (2 output neurons) is selected for the network model of 15–30–2, with input data normalized in the range 0.1–0.9. The nominal and interval class group data were converted to continuous values ranging between 0.1 and 0.9. Therefore, all the layers were normalized in the range 0.1–0.9. The categorical data and their interval class group were converted to a continuous values ranging 0.1–0.9. In this way, the continuous values became nominal for back-propagation modelling. The learning rate was set to 0.01, and the initial weights were randomly selected between 0.1 and 0.9. The MLP trained back-propagation algorithm was used to minimize the error between the predicted output values and the calculated output values. The algorithm propagated the error backwards iteratively by adjusting the weights. The number of epochs was set up to 10,000 and the root mean square error (RMSE) value used for the stopping criterion was set to 0.01. Most of the training data sets met the 0.011 RMSE goal. However, the maximum number of iterations was set to 10,000 epochs. When the latter case occurred, then the maximum RMSE value was 0.173 which selected for this study. Finally, the landslide susceptibility maps were generated and classified as very high, high, moderate, low, and very low susceptibility classes.

5.3 Result and Discussion

The prepared landslide susceptibility zonation map using ANN models classified the whole Darjeeling Himalaya into five, i.e. very low, low, moderate, high, and very high landslide susceptibility (Fig. 5.2). The results depicted that high landslide susceptibility zones occupied large areas of Darjeeling Himalaya which was followed by moderate, very high, low, and very low landslide susceptibility (Table 5.1). The study also revealed that very high landslide susceptible zones were mostly affected by landslide activities and very low landslide susceptible zones were being affected less by landslide phenomena. The frequency ratio value of very low, low, moderate, high, and very high landslide susceptible zones are 0.13, 0.33, 0.55, 0.87, and 2.32, respectively (Table 5.1 and Fig. 5.3). The prepared landslide susceptibility zonation map of Darjeeling Himalaya was validated with the help of derived ROC curve which revealed the accuracy level of 81.5% (Fig. 5.4). The study showed that the application of ANN in landslide susceptibility zonation mapping can be of immense important in regional planning and development of Darjeeling Himalaya as the model introduced a reasonable accuracy.

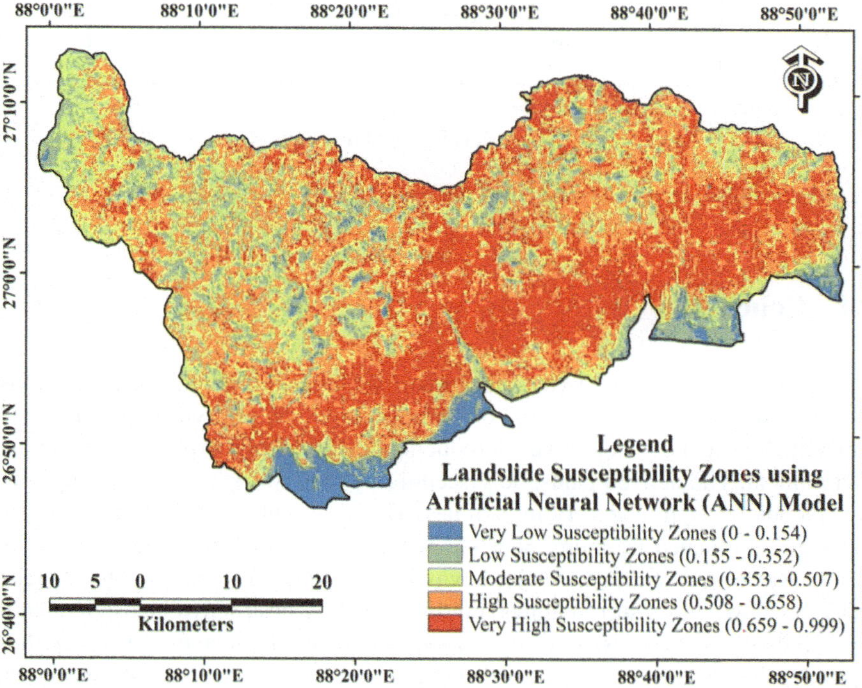

Fig. 5.2 Landslide susceptibility zonation map of Darjeeling Himalaya using ANN model

Table 5.1 Landslide susceptibility and frequency ratio

Landslide susceptibility zones	Total pixels	% of total pixels	Landslide pixels	% of landslide pixels	frequency ratio
Very low	108,230	4.20	26	0.53	0.13
Low	318,747	12.36	201	4.12	0.33
Moderate	632,113	24.52	662	13.56	0.55
High	980,817	38.04	1624	33.27	0.87
Very high	538,219	20.88	2369	48.53	2.32

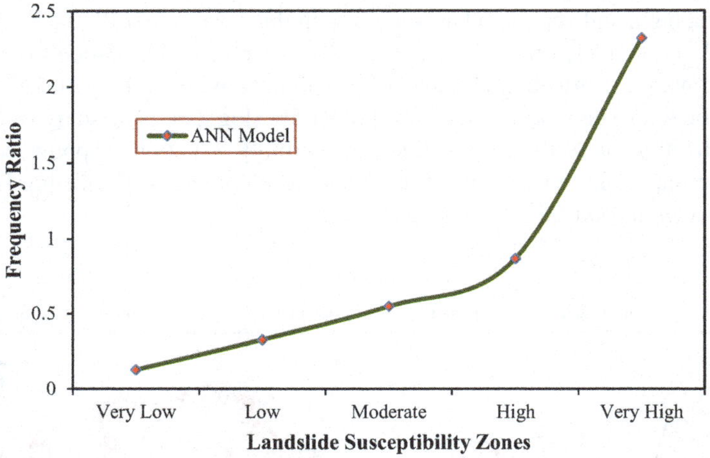

Fig. 5.3 Landslide susceptibility and Frequency ratio

5.4 Conclusion

The developed landslide susceptibility map of Darjeeling Himalaya using ANN showed a reasonable accuracy, and this product could be utilized by planners and policy-makers and developers for introducing developmental projects in mountain environment. The derived landslide susceptibility map was grouped into very low, low, moderate, high, and very high zones. The very high landslide susceptibility zones are dominated by more landslide occurrences where much attention to be paid. It was observed that the major towns such as Darjeeling, Kalimpong, Mirrik, and Kurseong and its surroundings are dominated by destructive landslide activities. All these locations exist mostly in moderate to very high landslide susceptibility zones. The weak lithology, concentration of drainage over the space, human intervention, high intensity rainfall, expansion of human settlement, and associated changes of land use and land cover played an important role in causing landslides in Darjeeling Himalaya. To mitigate and check landslides in Darjeeling Himalaya, site-specific management strategies are to be taken into account considering the

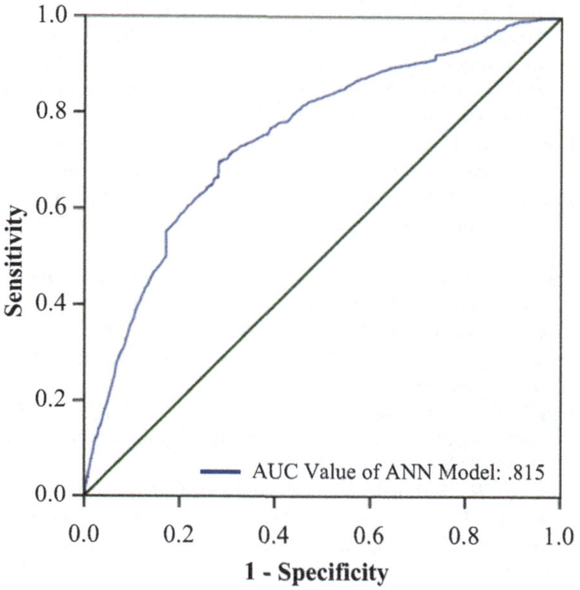

Fig. 5.4 Accuracy result using ROC curve

reared landslide susceptibility zonation. So, to find out the spatial distribution of landslide susceptibility and to introduce management options, ANN method is a significant statistical approach in the present day landslide science.

References

Aleotti, P. (2004). A warning system of rainfall-induced shallow failure. *Engineering Geology, 73*, 247–265.

Banerji, P. K., Guha, P. K., & Dhiman, L. C. (1980). Inverted metamorphism in the Sikkim-Darjeeling Himalaya. *Geological Society of India, 21*, 330–342.

Benardos, A. G., & Kaliampakos, D. C. (2004). A methodology for assessing geotechnical hazards for TBM tunnelling—illustrated by Athens Metro, Greece. *International Journal of Rock Mechanics and Mining Sciences, 41*(4), 987–999.

Benardos, A. G., Benardos, A. P. (2005). Applications of artificial neural networks in geotechnology. *Chron Min Metall, 15*(1), 65–81.

Bui, D. T., et al. (2012). Spatial prediction of landslide hazards in HoaBinh province (Vietnam): A comparative assessment of the efficacy of evidential belief functions and fuzzy logic models. *Catena, 96*, 28–40.

Caparrini, F., Caporali, E., & Castelli, F. (1996). Neural network analysis of satellite images for land cover discrimination. In P. Claps & F. Siccardi (Eds.), *Mediterranean Storms. Proceedings of the EGS Plinius Conference held at Maratea, Italy, October, 1999* (pp. 603–613).

Carrara, A., Cardinali, M., Detti, R., Guzzetti, F., Pasqui, V., & Reichenbach, P. (1991). GIS techniques and statistical models in evaluating landslide hazard. *Earth Surface Processes and Landforms, 16*, 427–445.

Carrara, A., Cardinali, M., Guazzetti, F., & Reichenbach, P. (1995). GIS techniques in mapping landslide hazard. In A. Carrara & F. Guzzetti (Eds.), *Geographical information systems in assessing natural hazards* (pp. 135–175). The Netherlands: Kluwer Academic Publishers.

Crosta, G. (1998). Regionalization of rainfall threshold: an aid to landslide hazard evaluation. *Environmental Geology, 35*(2–3), 131–145.

Crozier, M. J. (1999). Prediction of rainfall-triggered landslides: a test of the antecedent water status model. *Earth Surface Processes and Landforms, 24*, 825–833.

Emami, S. M. R., Iwao, Y., & Harada, T. (1998) Performance of Artificial Network system in prediction issues of earthquake engineering. *Proceedings of 8th International IAEG Congress* (pp. 733–738).

Fausett, L. (1994). *Fundamentals of neural networks architectures, Algorithms and applications* (p. 461). Upper Saddle River, NJ: Prentice Hall.

Froehlich, W., Gil, E., Kasza, I., & Starkel, L. (1990). Thresholds in the transformation of slopes and river channels in the Darjeeling Himalaya, India. *Mountain Research and Development, 10*(4), 301–312.

Glade, T., Crozier, M., & Smith, P. (2000). Applying probability determination to refine landslidetriggering rainfall thresholds using an empirical Antecedent Daily Rainfall Model. *Pure and Applied Geophysics, 157*, 1059–1079.

Guzzetti, F., Cardinali, M., Reichenbach, P., Cipolla, F., Sebastiani, C., Galli, M., & Salvati, P. (2004). Landslides triggered by the 23 November 2000 rainfall event in the Imperia Province, Western Liguria, Italy. *Engineering Geology, 73*, 229–245.

Guzzetti, F., Peruccacci, S., Rossi, M., & Stark, C. P. (2007) Rainfall thresholds for the initiation of landslides in central and southern Europe. *Meteorology and Atmospheric Physics*, online first version. DOI:https://doi.org/10.1007/s00703-007-0262-7. Retrieved July 8, 2007.

Hagan, T. M., Demuth, B. H., & Beale, H. M. (1996). *Neural network design, Electrical Engineering Series* (p. 730). Boston: Brooks/Cole.

Haykin, S. (1999). *Neural networks. A comprehensive foundation* (2nd ed.p. 696). Englewood Cliffs, NJ: Prentice Hall.

Hong et al. (2005). The influence of intense rainfall on the activity of large scale crystalline schist landslides in Shikoku Island, Japan. *Landslides, 2*, 97–105.

Hong, H., et al. (2015). Spatial prediction of landslide hazard at the Yihuang area (China) using two-class kernel logistic regression, alternating decision tree and support vector machines. *Catena, 133*, 266–281.

Ilanloo, M. (2011). A comparative study of fuzzy logic approach for landslide susceptibility mapping using GIS: an experience of Karaj dam basin in Iran. *Procedia - Social and Behavioral Sciences, 19*, 668–676.

Lee, S., Ryu, J., Min, K., & Won, J. (2001). *Proceedings of the Geoscience and Remote Sensing Symposium, IGARSS '01* (vol. 5, pp. 2364–2366), IEEE 2001 International.

Lee, S., Ryu, J. H., Lee, M. J., & Won, J. S. (2003a). Landslide susceptibility analysis using artificial neural network at Boun, Korea, *Environmental Geology, 44*, 820–833.

Lee, S., Ryu, J. H., Min, K. D., & Won, J. S. (2003b). Landslide susceptibility analysis using GIS and artificial neural network. *Earth Surface Processes and Landforms, 27*, 1361–1376.

Lee, S., Ryu, J. H., Lee, M. J., & Won, J. S. (2006). The Application of Artificial Neural Networks to Landslide Susceptibility Mapping at Janghung, Korea. *Mathematical Geology, 38*(2), 199–220.

Mayoraz, F., Cornu, T., & Vuillet, L. (1996) Using neural networks to predict slope movements. *Proceedings of the VII International Symposium on Landslides, Trondheim, June 1966, 1* (pp. 295–300). Rotterdam: Balkema.

McClelland, J. L., Rumelhart, D. E., & the PDP Research Group. (1986). *Parallel distributed processing: Explorations in the microstructure of cognition* (Vol. II). Cambridge, MA: MIT Press.

Montgomery, D. R., & Dietrich, W. E. (1994). A physically based model for the topographic control on shallow landsliding. *Water Resources Research, 30*(4), 1153–1171.

Pistocchi, A., Luzi, L., & Napolitano, P. (2002). The use of predictive modeling techniques for optimal exploitation of spatial databases: a case study in landslide hazard mapping with expert system-like methods. *Environmental Geology, 41*, 765–775.

Pradhan, B., & Lee, S. (2010a). Delineation of landslide hazard areas on Penang Island, Malaysia, by using frequency ratio, logistic regression, and artificial neural network models. *Environment and Earth Science, 60*, 1037–1054.

Pradhan, B., & Lee, S. (2010b). Landslide susceptibility assessment and factor effect analysis: backpropagation artificial neural networks and their comparison with frequency ratio and bivariate logistic regression modeling. *Environmental Modelling & Software, 25*(6), 747–759.

Pradhan, B., & Lee, S. (2010c). Regional landslide susceptibility analysis using back-propagation neural network model at Cameron Highland, Malaysia. *Landslides, 7*(1), 13–30.

Sezer, E. A., Pradhan, B., & Gokceoglu, C. (2011). Manifestation of an adaptive neuro-fuzzy model on landslide susceptibility mapping: Klang valley, Malaysia. *Expert Systems with Applications, 38*(7), 8208–8219.

Starkel, L. (1972). The role of catastrophic rainfall in the shaping of the relief of the Lower Himalaya (Darjeeling Hills). *Geographia Polonica, 21*, 103–147.

Terlien, M. T. J. (1997). Hydrological landslide triggering in ash covered slopes of Manizales (Columbia). *Geomorphology, 20*, 165–175.

Terlien, M. T. J. (1998). The determination of statistical and deterministic hydrological landslide triggering thresholds. *Environmental Geology, 35*(2–3), 124–130.

Tsangaratos, P., & Ilia, I. (2016). Landslide susceptibility mapping using a modified decision tree classifier in the Xanthi Perfection, Greece. *Landslides, 13*(2), 305–320.

Wieczorek, G. F. (1987). Effect of rainfall intensity and duration on debris flows in central Santa Cruz Mountains, California. In G. Crosta & G. F. Wieczorek (Eds.), *Debris flows/ avalanches: Processes, recognition and mitigation*, *Reviews in Engineering Geology* (Vol. 7, pp. 23–104). Boulder, CO: Geological Society of America.

Wieczorek, G. F. (1996). Landslide triggering mechanisms. In: Turner, A. K., Schuster, R. L. (Eds.), *Landslides: Investigation and mitigation* (pp. 76–79), Transportation Research Board, Special Report 247. Washington: National Research Council.

Wilson, R. C., & Wieczorek, G. F. (1995). Rainfall threshold for the initiation of debris flow at La Honda, California. *Environmental and Engineering Geoscience, 1*(1), 11–27.

Wu, X., Ren, F., & Niu, R. (2014). Landslide susceptibility assessment using object mapping units, decision tree, and support vector machine models in the Three Gorges of China. *Environmental Earth Sciences, 71*(11), 4725–4738.

Zezere, J. L., Trigo, R. M., & Trigo, I. F. (2005). Shallow and deep landslides induced by rainfall in the Lisbon region (Portugal): Assessment of relationships with the North Atlantic Oscillation. *Natural Hazards and Earth System Sciences, 5*, 331–344.

Zhou, W. (1999). Verfication of the nonparametric characteristics of backpropagation neural networks for image classification, IEEE Trans. *On Geoscience and Remote Sensing, 38*, 771–779.

Zhu, A.-X., et al. (2014). An expert knowledge-based approach to landslide susceptibility mapping using GIS and fuzzy logic. *Geomorphology, 214*, 128–138.

Chapter 6
Weighted Overlay Analysis (WOA) Model, Certainty Factor (CF) Model and Analytical Hierarchy Process (AHP) Model in Landslide Susceptibility Studies

6.1 Introduction

Landslides are the major environmental hazards which can be managed with the development of some concrete structures, increasing awareness amongst the people, identifying landslide-prone areas with the help of suitable techniques and introducing warning systems (Dai and Lee 2002). Varnes (1978, 1981) postulated that landslide susceptibility assessment aims to differentiate a land surface into homogeneous areas according to character of mass movement and propensity of failure. Jibson et al. (2000); Praise and Jibson (2000); Rautelal and Lakheraza (2000); Donati and Turrini (2002); and Zhou et al. (2002) suggested various techniques and methods for landslide susceptibility and risk assessment. Guzzetti et al. (1999) introduced suitable methods to delineate landslide hazard zones. In this chapter, weighted index overlay model, certainty factor approach, and analytical hierarchy process have been applied to assess landslide susceptibility of Darjeeling Himalaya considering 15 landslide conditioning factors, i.e. elevation, slope aspect, slope angle, slope curvature, geology, soil, lineament density, distance to lineament, drainage density, distance to drainage, stream power index (SPI), topographic wetted index (TWI), rainfall, normalized differential vegetation index (NDVI) and land use and land cover (LULC). All the techniques applied in landslide studies are quantitative as well as knowledge driven. Because, all the approaches are dealt with the assignment of the rating value to each class of all the factors and estimation of class weight as well as factor weight. In the *weighted index overlay model*, weighted values are assigned to all the data layers based on relative importance of the factors and their classes which requires an adequate knowledge with regard to landslide phenomena corresponding to the region concerned and finally their integration is being made on GIS environment to prepare landslide susceptibility map. In the *certainty factor approach*, certainty factor is derived for each class of all the data layers considering landslide inventory. This approach was first introduced by Shortliffe and Buchanan (1975) and then by Heckeman (1986). The value of the certainty factor was derived after Biaghi

S. Mandal, S. Mondal, *Statistical Approaches for Landslide Susceptibility Assessment and Prediction*, https://doi.org/10.1007/978-3-319-93897-4_6

et al. (1998) in the present work. Then, a linear combination model was performed on GIS environment to generate landslide susceptibility map of Darjeeling Himalaya. The validation of each approach was made with the help of ROC curve and based on the values of area under curve accuracy level was also estimated. *Analytical hierarchy process* (AHP) is a quantitative approach which considered the logical assignment of values against each class and factor to derive class weight and factor weight. This method was first proposed by Saaty (1980, 1994), Saaty and Vargas (1991, 2001), and Komac (2006). Komac (2006), Yalcin (2008), and Mandal and Maiti (2011) applied analytical hierarchy process (AHP) to delineate landslide hazard zonation map of the mountain region using RS and GIS. In this approach, each class as well as each factor was rated assigning the values 1–9 and a matrix was developed to estimate class weight and factor weight. Finally, a linear algorithm was applied on GIS platform to derive landslide susceptibility index values and to prepare landslide susceptibility zonation map of Darjeeling Himalaya. The preparation of landslide susceptibility map has become one of the significant ways to estimate, manage, and mitigate the landslide hazard for a region (Kienholz 1978; Spiker and Gori 2000; Chau et al. 2004). The landslide susceptibility indicates the area which is likely to undergo landslides in the future (Moradi et al. 2012). In Darjeeling Himalaya, most of the landslides are structurally controlled, i.e. a deep-seated rock wedge within Daling metasediments, plane failure along foliation surface in sheared phyllonitic rocks, and toppling caused by steep-dipping southwest dipping joint plane within quartzo-feldspathic gneiss (https://www.itc.nl/pdf/newsevents/landslides/vanwestern presentation.pdf). Geological structure of Darjeeling Himalaya plays an important role in the growth and development of drainage network. Such development of drainage network and its continuous branching promotes in slope steepening process and causes slope failure. So, the proper understanding of geologic, geomorphic, and hydrologic parameters and their integration to identify suitable sites and to prepare landslide susceptibility zones for regional development and planning have become much more significant in the present day geographical research.

6.2 Materials and Methods

6.2.1 Determination of Landslide Triggering Factors

The common method to study the landslide conditioning factors is to use questionnaire (oral judgement) and empirical study of the landslide inside the watershed associated to intensive field works (Abramson et al. 1995). After intensive field investigation and in consultation with the previous landslide occurrences phenomena, 15 landslide conditioning factors such as elevation, slope aspect, slope angle, slope curvature, geology, soil, lineament density, distance to lineament, drainage density, distance to drainage, stream power index (SPI), topographic wetted index (TWI), rainfall, normalized differential vegetation index (NDVI) and land use and land cover (LULC) were taken into account and their hierarchical arrangement was made to prepare landslide susceptibility map of Darjeeling Himalaya (Fig. 6.1).

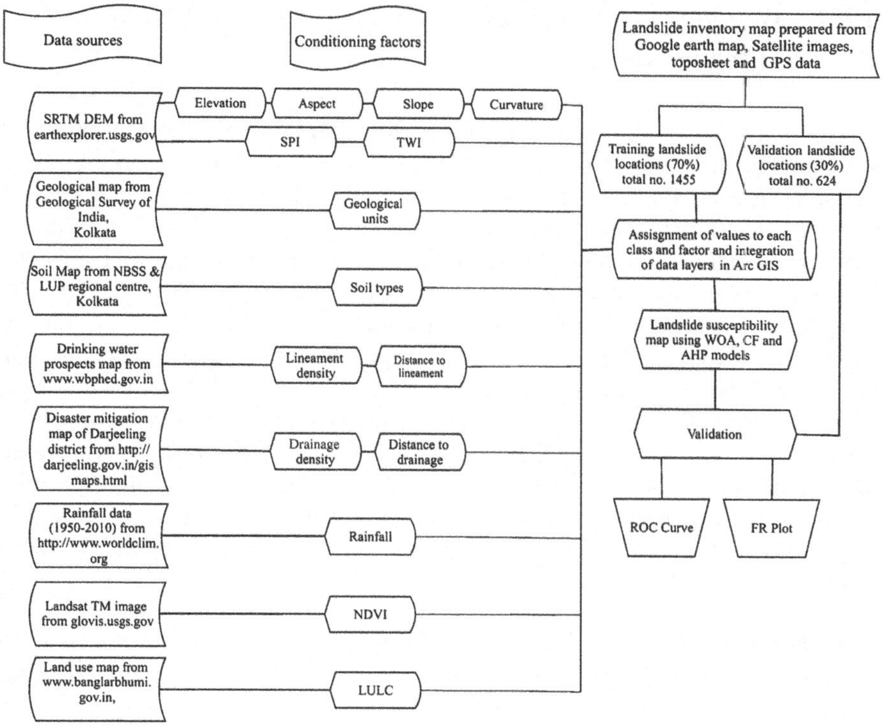

Fig. 6.1 Methodology of landslide susceptibility zonation map

6.2.2 *Generation of Various Thematic Data Layers*

SRTM DEM and GIS software played an important role in preparing data layers of elevation, slope angle, slope aspect, slope curvature, stream power index, and topographic wetness index of Darjeeling Himalaya. Lineament density and distance to lineament were digitized from drinking water prospects map. The distance to drainage and drainage density maps was developed preparing drainage vector layer from disaster mitigation map of Darjeeling district on GIS platform. Geological map was collected from Geological Survey of India (GSI, Kolkata) and the soil textural map based on soil taxonomy classification was prepared by NBSS (National Bureau of Soil Science, Kolkata) and LUP (Land Use Planning, Kolkata) Regional Centre, Kolkata where mainly nine soil types were identified, i.e. Tista river, W001, W002, W003, W004, W006, W007, W008 and W009, of the Darjeeling Himalaya. Rainfall is one of most important triggering factor of landslide susceptibility mapping. Rainfall map was prepared based on the data collected from the http://www.worldclim.org website and classified into ten groups accordingly. The details on the preparation of various thematic data layers with regard to landslide conditioning factors are described in Chap. 2.

6.2.3 Weighted Overlay Analysis (WOA) Model and Landslide Susceptibility

Preparation of landslide zonation map is an important technique which figure out spatial distribution of landslides and helps to introduce site-specific proper remedial measures in a rational way. In the present study, the interaction of different factors with the landslide inventory was studied separately and ultimately final coordination was made. For the preparation of landslide susceptibility zonation map of Darjeeling Himalaya ranking values to each class and parameters weight to each factor were assigned on the basis of the relative significance of each class. All the generated maps, i.e. elevation, slope aspect, slope angle, slope curvature, geology, soil, lineament density, distance to lineament, drainage density, distance to drainage, stream power index (SPI), topographic wetted index (TWI), rainfall, normalized differential vegetation index (NDVI) and land use and land cover (LULC) were reclassified with ranking values/weighted values (Table 6.1) depending on their degree of magnitude to landslide occurrences. Finally, a landslide susceptibility map was carved out applying *Weighted Overlay Analysis Model* using the following algorithm on GIS environment.

$$\textbf{Landslide Susceptibilty} = (\text{"Reclass\_LULC"} * 5) + (\text{"Reclass\_NDVI"} * 2.5) + (\text{"Reclass of Rainfall IDW"} * 14) + (\text{"Reclass of SPI"} * 2) + (\text{"Reclass of TWI"} * 1.5) + (\text{"Reclass\_D2D"} * 10) + (\text{"Reclass\_D2L"} * 3) + (\text{"Reclass of Lineament density"} * 4) + (\text{"Reclass of drainage density"} * 12) + (\text{"Reclass\_Soil"} * 7) + (\text{"Reclass\_Geology"} * 15) + (\text{"Reclass\_Curvature"} * 6) + (\text{"Reclass of Slope"} * 9) + (\text{"Reclass\_Aspect"} * 1) + (\text{"Reclass of elevation"} * 8) \tag{Eq. 6.1}$$

6.2.3.1 Process of Running Weighted Overlay Analysis (WOA) Model

The following steps have been taken into consideration for the preparation of the zonation map.

1. Selection of an evaluation scale: The evaluation scale represents the range of suitability. The values at one end of the scale represent one extreme of suitability and the values at the other end represent the other extreme.
2. Addition of input raster: This step includes the addition of input raster (data layers considered to prepare landslide susceptibility) to the weighted overlay table.
3. Setting of scale values: The cell values for each input raster/data layers in the analysis are assigned values from the evaluation scale. This makes it possible to perform arithmetic operations on raster that originally held dissimilar types of values.
4. Assignment of weights to input raster/data layers: Each input raster can be weighted or assigned a percentage influence, based on its importance. The total influence for all rasters must equal 100%.

Table 6.1 Modes of ranking of the intra subclass of parameters and distribution of priority-based weightage

Conditioning factor	Subclass	Subclass rank	Parameter weight/ eigen vector
Elevation (m)	130–470	1	8
	470–770	2	
	770–1100	3	
	1100–1300	4	
	1300–1600	5	
	1600–1900	6	
	1900–2200	7	
	2200–2600	8	
	2600–3000	9	
	3000–3600	10	
Aspect	Flat (−1)	1	1
	North (0–22.5)	3	
	Northeast (22.5–67.5)	9	
	East (67.5–112.5)	8	
	Southeast (112.5–157.5)	7	
	South (157.5–202.5)	6	
	Southwest (202.5–247.5)	10	
	West (247.5–292.5)	5	
	Northwest (292.5–337.5)	4	
	North (337.5–360)	2	
Slope (degree)	0–4.339	1	9
	4.339–8.677	2	
	8.677–12.515	3	
	12.515–15.853	4	
	15.853–19.024	5	
	19.024–22.027	6	
	22.027–25.031	7	
	25.031–28.202	8	
	28.202–32.373	9	
	32.373–42.553	10	
Curvature	−0.644 to −0.238	9	6
	−0.238 to −0.155	10	
	−0.155 to −0.095	8	
	−0.095 to −0.040	6	
	−0.040 to 0.015	1	
	0.015–0.064	3	
	0.064–0.120	2	
	0.120–0.185	4	
	0.185–0.267	5	
	0.267–0.756	7	

(continued)

Table 6.1 (continued)

Conditioning factor	Subclass	Subclass rank	Parameter weight/ eigen vector
Geology	Alluvium	1	15
	Baxa Series-Slates, Schists, Dolomite, Quartzites	4	
	Daling Series-Slates, Schists, Quartzites	8	
	Damuda (Gondwana)	10	
	Darjeeling Gneiss	6	
	Nahan Group (Tertiary) Limestone	3	
Soil	Tista River	1	7
	W001	9	
	W002	8	
	W003	6	
	W004	10	
	W006	4	
	W007	5	
	W008	7	
	W009	2	
Drainage density (km/sq. km)	0–0.694	1	12
	0.694–1.210	2	
	1.210–1.620	3	
	1.620–2.010	4	
	2.010–2.370	5	
	2.370–2.730	6	
	2.730–3.140	7	
	3.140–3.680	8	
	3.680–4.450	9	
	4.450–6.560	10	
Lineament density (km/sq. km)	0–0.056	1	4
	0. 056–0.168	2	
	0.168–0.287	3	
	0.287–0.399	4	
	0.399–0.505	5	
	0.505–0.610	6	
	0.610–0.716	7	
	0.716–0.835	8	
	0.835–0.966	9	
	0.966–1.590	10	

(continued)

Table 6.1 (continued)

Conditioning factor	Subclass	Subclass rank	Parameter weight/ eigen vector
Distance from lineament (m)	0–20	10	3
	20–50	9	
	50–100	8	
	100–400	7	
	400–1000	6	
	1000–2000	5	
	2000–3500	4	
	3500–5000	3	
	5000–6500	2	
	6500–8500	1	
Distance to drainage (m)	0–10	10	10
	10–30	9	
	30–70	8	
	70–150	7	
	150–300	6	
	300–700	5	
	700–1200	4	
	1200–1700	3	
	1700–2300	2	
	2300–3000	1	
SPI	−5.939 to −3.140	1	2
	−3.140 to −2.496	2	
	−2.496 to −2.013	3	
	−2.013 to −1.595	4	
	−1.595 to −1.209	5	
	−1.209 to −0.791	6	
	−0.791 to −0.373	7	
	−0.373 to 0.014	8	
	0.014 to 0.400	9	
	0.400 to 2.266	10	
TWI	−6907.755	1	1.5
	−6907.755 to −2849.717	2	
	−2849.717 to −1226.502	3	
	−1226.502 to −520.757	4	
	−520.757 to −167.884	5	
	−167.884 to −26.735	6	
	−26.735 to 43.840	7	
	43.840–255.564	8	
	255.564–714.298	9	
	714.298–2090.502	10	

(continued)

Table 6.1 (continued)

Conditioning factor	Subclass	Subclass rank	Parameter weight/ eigen vector
Mean annual rainfall (mm)	1076–1478	1	14
	1478–1798	2	
	1798–2097	3	
	2097–2355	4	
	2355–2551	5	
	2551–2705	6	
	2705–2860	7	
	2860–3056	8	
	3056–3293	9	
	3293–3705	10	
NDVI	−0.220 to 0.037	10	2.5
	0.037–0.110	9	
	0.110–0.166	8	
	0.166–0.214	7	
	0.214–0.263	6	
	0.263–0.315	5	
	0.315–0.368	4	
	0.368–0.427	3	
	0.427–0.493	2	
	0.493–0.667	1	
LULC	Social forestry	5	5
	Agricultural single crop	7	
	Forest	8	
	Rural settlement	4	
	Urban settlement	3	
	Agricultural plantation (tea)	6	
	Wasteland dry/Barren land	10	
	Public utility and facility	1	
	Wastelands with scrub	9	

5. Running the weighted overlay tool and extraction of landslide susceptibility map: The cell values of each input raster are multiplied by the raster's weight (or percent influence). The resulting cell values are added to produce the final output raster.

To perform the weighted index overlay model elevation, slope, aspect, curvature, lithology, distance to drainage, drainage density, distance to lineament, lineament density, soil, LULC, NDVI, SPI, TWI, and rainfall were assigned parameter weight/ eigen vector of 8, 9, 1, 10, 15, 10, 12, 3, 4, 7, 5, 2.5, 2, 1.5, and 14, respectively (Table 6.1), considering the relative importance of each factor to landslide. In the same way, each class of landslide causative factor was assigned a rank value to run the weighted index overlay model on GIS environment.

6.2.4 *Certainty Factor (CF) Model and Landslide Susceptibility*

Certainty factor approach, one of the most important methods which deals with the problem of combination of heterogeneous data. Kanungo et al. (2011), Sujatha et al. (2012), and Liu et al. (2014) widely used CF model for the mapping of landslide susceptibility zonation map. The certainty factor method was first proposed by Shortliffe and Buchanan (1975), and later it was improved by Heckeman (1986).

The method can be expressed with the help of following equation (eq. 6.2):

$$\mathrm{CF} = \begin{cases} \dfrac{\mathrm{ppa}-\mathrm{pps}}{\mathrm{ppa}(1-\mathrm{pps})} & \text{if } \mathrm{ppa} \geq \mathrm{pps} \\ \dfrac{\mathrm{ppa}-\mathrm{pps}}{\mathrm{ppa}(1-\mathrm{ppa})} & \text{if } \mathrm{ppa} < \mathrm{pps} \end{cases} \qquad \text{(Eq. 6.2)}$$

where CF is the certainty factor, ppa is the conditional probability of having a number of landslides in a class "a", and pps is the prior probability of having the total number of landslides in the study area "A".

The data layers were combined pair-wise according to the integration rules (Chung and Fabbri 1993; Biaghi et al. 1998). The combination of CF values of two thematic layers "z" is expressed in the following equation as given by Biaghi et al. (1998), Pourghasemi et al. (2013); Dou et al. (2014); Ilia et al. (2015):

$$Z = \begin{cases} x+y-xy & x, y \geq 0 \\ \dfrac{x+y}{1-\min\left(\left\|x\right\| \| \; y\right)} & x, y \text{ opposite } sign \\ x+y+xy, & x, y < 0 \end{cases} \qquad \text{(Eq. 6.3)}$$

The CF values of all the thematic data layers were derived in ArcGIS 10.1 and Microsoft Excel based on equation (6.2) (Table 6.2). Then, pair-wise CF values of the causative factor were estimated based on the combination rule (eq. 6.3). By using the integration rule of equation, the pair-wise combination was repeatedly performed until all the CF layers are combined to generate the landslide susceptibility.

6.2.5 *Analytical Hierarchy Process (AHP) Model and Landslide Susceptibility*

The analytical hierarchy process (AHP) is a semi-qualitative, multi-objective, and multi criteria decision making (MCDM) method for landslide susceptibility assessment which involves a matrix-based pair-wise comparison of the contribution of different factors and which supplies a flexible and easily understood way of

Table 6.2 Spatial relationship between each landslide conditioning factors and landslide by certainty factor model (CFM)

Conditioning factor	Class	Number of pixels in domain	Number of landslide events	ppa	pps	CF
Elevation (m)	130–470	409,787	263	0.000642	0.000564	0.12
	470–770	437,151	459	0.001050	0.000564	0.46
	770–1100	437,588	240	0.000548	0.000564	−0.03
	1100–1300	246,682	144	0.000584	0.000564	0.03
	1300–1600	342,538	119	0.000347	0.000564	−0.38
	1600–1900	260,070	104	0.000400	0.000564	−0.29
	1900–2200	171,257	70	0.000409	0.000564	−0.28
	2200–2600	142,581	32	0.000642	0.000564	−0.60
	2600–3000	80,867	11	0.001050	0.000564	−0.76
	3000–3600	49,605	13	0.000548	0.000564	−0.54
Aspect	Flat (−1)	231	0	0	0.000564	−1.00
	North (0–22.5)	136,124	40	0.000294	0.000564	−0.48
	Northeast (22.5–67.5)	296,730	139	0.000468	0.000564	−0.17
	East (67.5–112.5)	395,973	242	0.000611	0.000564	0.08
	Southeast (112.5–157.5)	413,694	319	0.000771	0.000564	0.27
	South (157.5–202.5)	386,071	278	0.00072	0.000564	0.22
	Southwest (202.5–247.5)	299,310	247	0.000825	0.000564	0.32
	West (247.5–292.5)	266,605	115	0.000431	0.000564	−0.24
	Northwest (292.5–337.5)	250,427	49	0.000196	0.000564	−0.65
	North (337.5–360)	132,961	26	0.000196	0.000564	−0.65
Slope (degree)	0–4.339	187,019	62	0.000332	0.000564	−0.41
	4.339–8.677	255,475	119	0.000466	0.000564	−0.17
	8.677–12.515	343,632	187	0.000544	0.000564	−0.04
	12.515–15.853	374,691	227	0.000606	0.000564	0.07
	15.853–19.024	427,124	220	0.000515	0.000564	−0.09
	19.024–22.027	382,230	215	0.000562	0.000564	0.00
	22.027–25.031	278,609	176	0.000632	0.000564	0.11
	25.031–28.202	189,918	127	0.000669	0.000564	0.16
	28.202–32.373	107,393	89	0.000829	0.000564	0.32
	32.373–42.553	32,035	33	0.00103	0.000564	0.45

(continued)

Table 6.2 (continued)

Conditioning factor	Class	Number of pixels in domain	Number of landslide events	ppa	pps	CF
Curvature	−0.644 to −0.238	48,330	69	0.001428	0.000564	0.60
	−0.238 to −0.155	168,472	166	0.000985	0.000564	0.43
	−0.155 to −0.095	286,017	218	0.000762	0.000564	0.26
	−0.095 to −0.040	390,224	262	0.000671	0.000564	0.16
	−0.040 to 0.015	577,690	222	0.000384	0.000564	−0.32
	0.015–0.064	406,708	163	0.000401	0.000564	−0.29
	0.064–0.120	330,604	155	0.000469	0.000564	−0.17
	0.120–0.185	234,280	110	0.00047	0.000564	−0.17
	0.185–0.267	108,702	65	0.000598	0.000564	0.06
	0.267–0.756	27,099	26	0.000959	0.000564	0.41
Geology	Alluvium	104,603	10	0.000096	0.000564	−0.83
	Baxa Series-Slates, Schists, Dolomite, Quartzites	2344	2	0.000853	0.000564	0.34
	Daling Series-Slates, Schists, Quartzites	712,203	556	0.000781	0.000564	0.28
	Damuda (Gondwana)	35,988	76	0.002112	0.000564	0.73
	Darjeeling Gneiss	1,523,340	634	0.000416	0.000564	−0.26
	Nahan Group (Tertiary) Limestone	199,648	177	0.000887	0.000564	0.36
Soil	Tista River	14,862	17	0.000564	0.000564	0.51
	W001	261,470	45	0.000564	0.000564	−0.70
	W002	1434,692	731	0.000564	0.000564	−0.10
	W003	221,855	167	0.000564	0.000564	0.25
	W004	446,612	431	0.000564	0.000564	0.42
	W006	75,842	2	0.000564	0.000564	−0.95
	W007	85,141	39	0.000564	0.000564	−0.19
	W008	35,338	23	0.000564	0.000564	0.13
	W009	2314	0	0.000564	0.000564	−1.00
Drainage density (km/sq. km)	0–0.694	243,293	80	0.000329	0.000564	−0.42
	0.694–1.210	409,926	164	0.0004	0.000564	−0.29
	1.210–1.620	469,031	290	0.000618	0.000564	0.09
	1.620–2.010	498,755	335	0.000672	0.000564	0.16
	2.010–2.370	393,587	258	0.000656	0.000564	0.14
	2.370–2.730	272,462	187	0.000686	0.000564	0.18
	2.730–3.140	166,766	94	0.000564	0.000564	0.00
	3.140–3.680	76,489	28	0.000366	0.000564	−0.35
	3.680–4.450	34,741	12	0.000345	0.000564	−0.39
	4.450–6.560	13,076	7	0.000535	0.000564	−0.05

(continued)

Table 6.2 (continued)

Conditioning factor	Class	Number of pixels in domain	Number of landslide events	ppa	pps	CF
Lineament density (km/sq. km)	0–0.056	1,309,367	613	0.000469	0.000564	−0.17
	0. 056–0.168	300,996	162	0.000538	0.000564	−0.05
	0.168–0.287	230,519	148	0.000642	0.000564	0.12
	0.287–0.399	192,603	128	0.000665	0.000564	0.15
	0.399–0.505	162,361	104	0.000641	0.000564	0.12
	0.505–0.610	141,605	109	0.00077	0.000564	0.27
	0.610–0.716	102,252	93	0.00091	0.000564	0.38
	0.716–0.835	73,150	62	0.000848	0.000564	0.33
	0.835–0.966	48,815	25	0.000512	0.000564	−0.09
	0.966–1.590	16,458	11	0.000668	0.000564	0.16
Distance from lineament (m)	0–20	23,691	63	0.002659	0.000564	0.79
	20–50	36,643	66	0.001801	0.000564	0.69
	50–100	63,438	101	0.001592	0.000564	0.65
	100–400	406,582	253	0.000622	0.000564	0.09
	400–1000	787,207	383	0.000487	0.000564	−0.14
	1000–2000	781,676	411	0.000526	0.000564	−0.07
	2000–3500	369,169	168	0.000455	0.000564	−0.19
	3500–5000	95,146	10	0.000105	0.000564	−0.81
	5000–6500	10,670	0	0	0.000564	−1.00
	6500–8500	3904	0	0	0.000564	−1.00
Distance to drainage (m)	0–10	92,856	129	0.001389	0.000564	0.59
	10–30	184,399	174	0.000944	0.000564	0.40
	30–70	356,416	234	0.000657	0.000564	0.14
	70–150	628,425	360	0.000573	0.000564	0.01
	150–300	763,840	365	0.000478	0.000564	−0.15
	300–700	498,808	174	0.000349	0.000564	−0.38
	700–1200	45,871	18	0.000392	0.000564	−0.30
	1200–1700	5659	1	0.000177	0.000564	−0.69
	1700–2300	1593	0	0	0.000564	−1.00
	2300–3000	259	0	0	0.000564	−1.00
SPI	−5.939 to −3.140	14,266	13	0.000911	0.000564	0.38
	−3.140 to −2.496	57,082	42	0.000736	0.000564	0.23
	−2.496 to −2.013	132,095	98	0.000742	0.000564	0.24
	−2.013 to −1.595	223,338	129	0.000578	0.000564	0.02
	−1.595 to −1.209	296,386	158	0.000533	0.000564	−0.06
	−1.209 to −0.791	361,250	175	0.000484	0.000564	−0.14
	−0.791 to −0.373	379,786	240	0.000632	0.000564	0.11
	−0.373 to 0.014	571,397	270	0.000473	0.000564	−0.16
	0.014–0.400	391,010	220	0.000563	0.000564	0.00
	0.400–2.266	151,516	110	0.000726	0.000564	0.22

(continued)

Table 6.2 (continued)

Conditioning factor	Class	Number of pixels in domain	Number of landslide events	ppa	pps	CF
TWI	−6907.755	232	0	0	0.000564	−1.00
	−6907.755 to −2849.717	848	0	0	0.000564	−1.00
	−2849.717 to −1226.502	2601	0	0	0.000564	−1.00
	−1226.502 to −520.757	6953	2	0.000288	0.000564	−0.49
	−520.757 to −167.884	36,026	12	0.000333	0.000564	−0.41
	−167.884 to −26.735	454,856	212	0.000466	0.000564	−0.17
	−26.735 to 43.840	1,987,528	1187	0.000597	0.000564	0.06
	43.840–255.564	76,829	38	0.000495	0.000564	−0.12
	255.564–714.298	10,964	4	0.000365	0.000564	−0.35
	714.298–2090.502	1289	0	0	0.000564	−1.00
Mean annual rainfall (mm)	1076–1478	54,494	9	0.000165	0.000564	−0.71
	1478–1798	70,178	21	0.000299	0.000564	−0.47
	1798–2097	73,731	9	0.000122	0.000564	−0.78
	2097–2355	88,710	30	0.000338	0.000564	−0.40
	2355–2551	268,934	144	0.000535	0.000564	−0.05
	2551–2705	551,234	241	0.000437	0.000564	−0.23
	2705–2860	608,233	333	0.000547	0.000564	−0.03
	2860–3056	423,558	337	0.000796	0.000564	0.29
	3056–3293	277,977	240	0.000863	0.000564	0.35
	3293–3705	161,077	91	0.000565	0.000564	0.00
NDVI	−0.220 to 0.037	34,131	64	0.001875	0.000564	0.70
	0.037–0.110	77,506	138	0.001781	0.000564	0.68
	0.110–0.166	167,231	177	0.001058	0.000564	0.47
	0.166–0.214	273,220	223	0.000816	0.000564	0.31
	0.214–0.263	381,618	249	0.000652	0.000564	0.14
	0.263–0.315	451,331	196	0.000434	0.000564	−0.23
	0.315–0.368	428,060	172	0.000402	0.000564	−0.29
	0.368–0.427	364,538	120	0.000329	0.000564	−0.42
	0.427–0.493	257,616	83	0.000322	0.000564	−0.43
	0.493–0.667	142,875	33	0.000231	0.000564	−0.59

(continued)

Table 6.2 (continued)

Conditioning factor	Class	Number of pixels in domain	Number of landslide events	ppa	pps	CF
LULC	Social forestry	309,800	148	0.000478	0.000564	−0.15
	Agricultural single crop	331,856	199	0.0006	0.000564	0.06
	Forest	1,430,826	902	0.00063	0.000564	0.10
	Rural settlement	126,193	41	0.000325	0.000564	−0.42
	Urban settlement	16,672	2	0.00012	0.000564	−0.79
	Agricultural plantation (tea)	337,588	133	0.000394	0.000564	−0.30
	Wasteland dry/Barren land	399	2	0.005013	0.000564	0.89
	Public utility and facility	3052	0	0	0.000564	−1.00
	Wastelands with scrub	21,740	28	0.001288	0.000564	0.56

analysing complicated problems (Saaty 1980). This process enables the user to arrive at a scale of preferences drawn from a set of alternatives. The AHP method widely helped the planner, policy, and decision-makers to find out their best suitable goal in site selection, suitability analysis, routing modelling, regional planning, and landslide susceptibility analysis (Ayalew et al. 2005), and it is a problem-solving methodical process for representing the elements of any problem (Saaty and Vargas 1991). For the application of this model, it is necessary to break a complex unstructured problem down into its component factors which are the factors chosen in this study, arrange these factors in a hierarchical order, assign numerical values according to their subjective judgements on the relative importance of each factor, and synthesize the judgements to determine the priorities to be assigned to these factors (Saaty and Vargas 2001). When arranging the factors in a hierarchical order, there should be relative importance of one parameter to another forming a pair-wise comparison matrix with scores given in Table 6.3. In the formulation of a pair-wise comparison matrix, each factor is rated against every other factor by assigning a relative dominant value between 1 and 9 (Table 6.3). In the matrix, the factor on the vertical axis is more important than the factor on the horizontal axis, and this value varies between 1 and 9. Conversely, the value varies between the reciprocals 0.5 (1/2) and 0.11 (1/9). In these techniques, at first, the effects of each parameter to the susceptibility of landslides relative to each other were determined by dual evaluation in determining the preferences in the effects of the parameters to the landslide susceptibility map. Generally, the determination of the values of the parameters relative to each other is a situation that is dependent on the choices of the decision-maker. However, in this study, both the comparison of the parameters relative to each other and the determination of the decision alternatives, namely the effect values of the sub-criteria of the factors weights (priority) were obtained based on

Table 6.3 The fundamental scale of preference between two parameters for pair-wise comparison matrix in AHP (Saaty 1980)

Scales or importance rank	Degree of preference	Explanation
1	Equal importance	Two criteria or activities contribute equally to the objective
3	Moderate important to one over another	Experience and judgement slightly to moderately favour one criteria over another
5	Strongly important	Experience and judgements strongly or essentially favour one criteria over another
7	Very strongly important	Judgements and experience is strongly favoured over another and its dominance is showed in practice
9	Extremely important	The evidence favouring one criteria over another is of the highest degree possible of an affirmation
2, 4, 6, and 8	Intermediate values between two adjacent judgements	Used to represent compromises between the preferences in weights 1, 3, 5, 7, and 9
Reciprocals	Opposites	Used for inverse comparison

evaluation of landslide inventory maps (Basu and Ghatowar 1988). In AHP, class weights (Priority) and factor weights (priority) with reasonable consistency ratio (CR) were derived to integrate all the data layers as well as to prepare landslide susceptibility map of the Darjeeling Himalaya (Tables 6.4 and 6.5).

In the AHP model, an index of consistency, known as the consistency ratio (CR), is used to indicate the probability that the matrix judgements were randomly generated (Saaty 1977, 1980, 1994).

$$CR = \left(\frac{CI}{RI}\right) \quad \text{(Eq. 6.4)}$$

where RI is the average of the resulting consistency index depending on the order of the matrix given by Saaty (1977) and CI is the consistency index (Table 6.6) and can be expressed as

$$CI = \{(\lambda_{max} - n) / (n - 1)\} \quad \text{(Eq. 6.5)}$$

where λ_{max} is the largest or principal eigenvalue of the matrix and can be easily calculated from the matrix and n is the order of the matrix.

The consistency ratio (CR) is a ratio between the matrix's consistency index and random index, and it is ranges from 0 to 1.CR value of 0.1 or less indicates a logical level of consistency (Malczewski 1999) and CR value greater than 0.1 indicate the models required as revision of the judgement in the matrix due to an inconsistent treatment for particular factor ratings.

Table 6.4 Pair-wise comparison matrix of landslide conditioning factors

	Geology	Rainfall	Drainage density	Distance to drainage	Slope	Elevation	Soil	Curvature	LULC	Lineament density	Distance to lineament	NDVI	SPI	TWI	Aspect	Eigen vector	Consistency ratio
Geology	1	1	2	2	2	3	3	4	4	5	5	6	6	7	8	0.182	0.01607172
Rainfall	1	1	1	2	2	3	3	4	5	6	6	7	7	8	9	0.167	
Drainage density	1/2	1	1	2	2	2	3	3	4	5	6	7	7	8	8	0.141	
Distance to drainage	1/2	1/2	½	1	1	1	1	2	2	3	4	5	5	6	7	0.082	
Slope	1/2	1/2	½	1	1	1	1	2	2	3	3	4	4	5	5	0.079	
elevation	1/3	1/3	½	1	1	1	1	2	2	3	4	5	5	6	6	0.071	
Soil	1/3	1/3	1/3	1	1	1	1	1	1	2	3	3	4	4	5	0.061	
Curvature	1/4	1/4	1/3	1/2	1/2	1/2	1	1	1	2	2	2	2	3	3	0.044	
LULC	1/4	1/5	¼	1/2	1/2	1/2	1	1	1	2	2	2	3	4	4	0.042	
Lineament density	1/5	1/6	1/5	1/3	1/3	1/3	1/2	1/2	1/2	1	1	1	2	3	3	0.029	
Distance to lineament	1/5	1/6	1/6	1/4	1/3	1/4	1/3	1/2	1/2	1	1	1	1	2	2	0.026	
NDVI	1/6	1/7	1/7	1/5	1/4	1/5	1/3	1/2	1/2	1	1	1	1	1	2	0.022	
SPI	1/6	1/7	1/7	1/5	1/4	1/5	1/4	1/2	1/3	1/2	1	1	1	1	1	0.021	
TWI	1/7	1/8	1/8	1/6	1/5	1/6	1/4	1/3	1/4	1/3	1/2	1	1	1	1	0.018	
Aspect	1/8	1/9	1/8	1/7	1/5	1/6	1/5	1/3	1/4	1/3	1/2	1/2	1	1	1	0.016	

Table 6.5 Weightages for different parameters and subclass ranking pattern

Conditioning factor	Subclass	Subclass rank	Parameter weight/ eigen vector
Elevation (m)	130–470	1	0.071
	470–770	2	
	770–1100	3	
	1100–1300	4	
	1300–1600	5	
	1600–1900	6	
	1900–2200	7	
	2200–2600	8	
	2600–3000	9	
	3000–3600	10	
Aspect	Flat (−1)	1	0.016
	North (0–22.5)	3	
	Northeast (22.5–67.5)	9	
	East (67.5–112.5)	8	
	Southeast (112.5–157.5)	7	
	South (157.5–202.5)	6	
	Southwest (202.5–247.5)	10	
	West (247.5–292.5)	5	
	Northwest (292.5–337.5)	4	
	North (337.5–360)	2	
Slope (degree)	0–4.339	1	0.079
	4.339–8.677	2	
	8.677–12.515	3	
	12.515–15.853	4	
	15.853–19.024	5	
	19.024–22.027	6	
	22.027–25.031	7	
	25.031–28.202	8	
	28.202–32.373	9	
	32.373–42.553	10	
Curvature	−0.644 to −0.238	9	0.044
	−0.238 to −0.155	10	
	−0.155 to −0.095	8	
	−0.095 to −0.040	6	
	−0.040 to 0.015	1	
	0.015–0.064	3	
	0.064–0.120	2	
	0.120–0.185	4	
	0.185–0.267	5	
	0.267–0.756	7	

(continued)

Table 6.5 (continued)

Conditioning factor	Subclass	Subclass rank	Parameter weight/ eigen vector
Geology	Alluvium	1	0.182
	Baxa Series-Slates, Schists, Dolomite, Quartzites	4	
	Daling Series-Slates, Schists, Quartzites	8	
	Damuda (Gondwana)	10	
	Darjeeling Gneiss	6	
	Nahan Group (Tertiary) Limestone	3	
Soil	Tista River	1	0.061
	W001	9	
	W002	8	
	W003	6	
	W004	10	
	W006	4	
	W007	5	
	W008	7	
	W009	2	
Drainage density (km/ sq. km)	0–0.694	1	0.141
	0.694–1.210	2	
	1.210–1.620	3	
	1.620–2.010	4	
	2.010–2.370	5	
	2.370–2.730	6	
	2.730–3.140	7	
	3.140–3.680	8	
	3.680–4.450	9	
	4.450–6.560	10	
Lineament density (km/ sq. km)	0–0.056	1	0.029
	0.056–0.168	2	
	0.168–0.287	3	
	0.287– 0.399	4	
	0.399–0.505	5	
	0.505–0.610	6	
	0.610–0.716	7	
	0.716–0.835	8	
	0.835–0.966	9	
	0.966–1.590	10	

(continued)

Table 6.5 (continued)

Conditioning factor	Subclass	Subclass rank	Parameter weight/ eigen vector
Distance from lineament (m)	0–20	10	0.026
	20–50	9	
	50–100	8	
	100–400	7	
	400–1000	6	
	1000–2000	5	
	2000–3500	4	
	3500–5000	3	
	5000–6500	2	
	6500–8500	1	
Distance to drainage (m)	0–10	10	0.082
	10–30	9	
	30–70	8	
	70–150	7	
	150–300	6	
	300–700	5	
	700–1200	4	
	1200–1700	3	
	1700–2300	2	
	2300–3000	1	
SPI	−5.939 to −3.140	1	0.021
	−3.140 to −2.496	2	
	−2.496 to −2.013	3	
	−2.013 to −1.595	4	
	−1.595 to −1.209	5	
	−1.209 to −0.791	6	
	−0.791 to −0.373	7	
	−0.373 to 0.014	8	
	0.014–0.400	9	
	0.400–2.266	10	
TWI	−6907.755	1	0.018
	−6907.755 to −2849.717	2	
	−2849.717 to −1226.502	3	
	−1226.502 to −520.757	4	
	−520.757 to −167.884	5	
	−167.884 to −26.735	6	
	−26.735 to 43.840	7	
	43.840–255.564	8	
	255.564–714.298	9	
	714.298–2090.502	10	

(continued)

Table 6.5 (continued)

Conditioning factor	Subclass	Subclass rank	Parameter weight/ eigen vector
Mean annual rainfall (mm)	1076–1478	1	0.167
	1478–1798	2	
	1798–2097	3	
	2097–2355	4	
	2355–2551	5	
	2551–2705	6	
	2705–2860	7	
	2860–3056	8	
	3056–3293	9	
	3293–3705	10	
NDVI	−0.220 to 0.037	10	0.022
	0.037–0.110	9	
	0.110–0.166	8	
	0.166–0.214	7	
	0.214–0.263	6	
	0.263–0.315	5	
	0.315–0.368	4	
	0.368–0.427	3	
	0.427–0.493	2	
	0.493–0.667	1	
LULC	Social forestry	5	0.042
	Agricultural single crop	7	
	Forest	8	
	Rural settlement	4	
	Urban settlement	3	
	Agricultural plantation (tea)	6	
	Wasteland dry/Barren land	10	
	Public utility and facility	1	
	Wastelands with scrub	9	

Table 6.6 Random consistency index (RI) (Saaty 1977)

N	1	2	3	4	5	6	7	8	9	10
RI	0	0	0.58	0.09	1.12	1.32	1.41	1.45	1.49	1.49

The developed pair-wise matrix and estimated eigen vector/prioritized factor rating value of all the landslide causative factors, i.e. geology, rainfall, drainage density, distance to drainage, slope, elevation, soil, curvature, LULC, lineament density, distance to lineament, NDVI, SPI, TWI, and slope aspect are 0.182, 0.167, 0.141, 0.082, 0.079, 0.071, 0.061, 0.044, 0.042, 0.029, 0.026, 0.022, 0.021, 0.018, and 0.016, respectively. The developed matrix represents acceptable consistency ratio value of 0.016.

To perform AHP and to prepare landslide susceptibility zonation map of Darjeeling Himalaya the following algorithm was applied on GIS environment (eq. 6.6).

$$\textbf{Landslide Susceptibility} = (\text{"Reclass\_LULC"} * 0.042) + (\text{"Reclass\_NDVI"} * 0.022)$$
$$+ (\text{"Reclass of Rainfall IDW"} * 0.167) + (\text{"Reclass of SPI"} * 0.021)$$
$$+ (\text{"Reclass of TWI"} * 0.018) + (\text{"Reclass\_D2D"} * 0.082)$$
$$+ (\text{"Reclass\_D2L"} * 0.026) + (\text{"Reclass of Lineament density"} * 0.029)$$
$$+ (\text{"Reclass of drainage density"} * 0.141) + (\text{"Reclass\_Soil"} * 0.079)$$
$$+ (\text{"Reclass\_Geology"} * 0.182) + (\text{"Reclass\_Curvature"} * 0.044)$$
$$+ (\text{"Reclass of Slope"} * 0.079) + (\text{"Reclass\_Aspect"} * 0.016) + (\text{"Reclass of elevation"} * 0.071) \quad \text{(Eq. 6.6)}$$

6.3 Result and Discussion

6.3.1 Weighted Overlay Analysis (WOA) Model and Landslide Susceptibility

Darjeeling Himalaya was divided into five landslide susceptibility zones based on *weighted index overlay model,* i.e. very low, low, moderate, high, and very high (Fig. 6.2). With regard to areal coverage moderate landslide susceptibility (31.74%) ranks first which is followed by high (31.29%), low (16.42%), very high (16.37%), and very low (4.18%). High and very high landslide susceptibility zones are dominated by landslides where the occurrences of landslide phenomena and the probability are very high (Table 6.7). There is a positive relationship between landslide susceptibility and frequency ratio which depicted high landslide susceptibility zones having maximum frequency ration and greater probability of landslides.

6.3.2 Certainty Factor (CF) Model and Landslide Susceptibility

The application of *certainty factor approach* and derived landslide susceptibility map of Darjeeling Himalaya was also classified into five such as very low, low, moderate, high, and very high landslide susceptibility zones (Fig. 6.3). This approach showed that Darjeeling Himalaya is dominated by moderate to high landslide susceptibility. It was found that high and very high landslide susceptibility zones were being affected by landslide events. The frequency ratio values of very

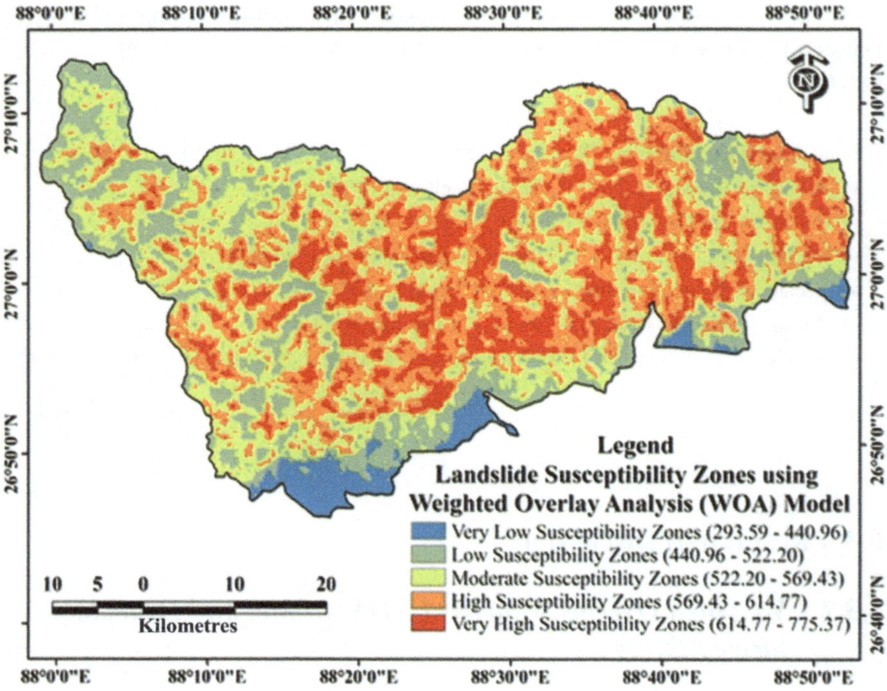

Fig. 6.2 Landslide susceptibility zonation map using WOA model

low, low, moderate, high, and very high landslide susceptibility zones are 0.27, 0.36, 0.81, 1.25, and 2.89, respectively, which showed high landslide susceptibility is attributed with greater probability of landslides over the space (Table 6.7).

6.3.3 *Analytical Hierarchy Process (AHP) Model and Landslide Susceptibility*

Darjeeling Himalaya was classified into five landslide susceptibility zones such as very low, low, moderate, high, and very high (Fig. 6.4) using analytical hierarchy process (AHP). The prepared landslide susceptibility map based on AHP revealed that Darjeeling Himalaya is experienced with 31.40% of moderate landslide susceptibility, 31.28% of high landslide susceptibility, 16.37% of high landslide susceptibility, and 20.655% of low to very low landslide susceptibility. The relationship between landslide susceptibility map based on AHP and landslide inventory showed that high and very high landslide susceptibility zones are highly affected by landslide activities where frequency ratio value and landslide probability is very high (Table 6.7).

Table 6.7 Area and frequency ratio value of different susceptibility classes using AHP, WOA, and CF Models

Landslide susceptibility zones	Total pixels			% of Total pixels			Landslide pixels			% of Landslide pixels			Frequency ratio		
	Using AHP model	Using WOM model	Using CF model	Using AHP model	Using WOM model	Using CF model	Using AHP model	Using WOM model	Using CF model	Using AHP model	Using WOM model	Using CF model	Using AHP model	Using WOM model	Using CF model
Very low	105,750	107,763	260,389	4.10	4.18	10.10	2	2	131	0.04	0.04	2.68	0.01	0.01	0.27
Low	426,768	42,3280	570,315	16.55	16.42	22.12	318	340	386	6.51	6.97	7.91	0.39	0.42	0.36
Moderate	809,547	818,403	798,599	31.40	31.74	30.98	1097	1015	1226	22.47	20.79	25.11	0.72	0.66	0.81
High	806,489	806,617	661,118	31.28	31.29	25.64	1853	1896	1567	37.96	38.83	32.09	1.21	1.24	1.25
Very high	429,572	422,063	287,705	16.67	16.37	11.16	1612	1629	1572	33.02	33.38	32.20	1.98	2.04	2.89

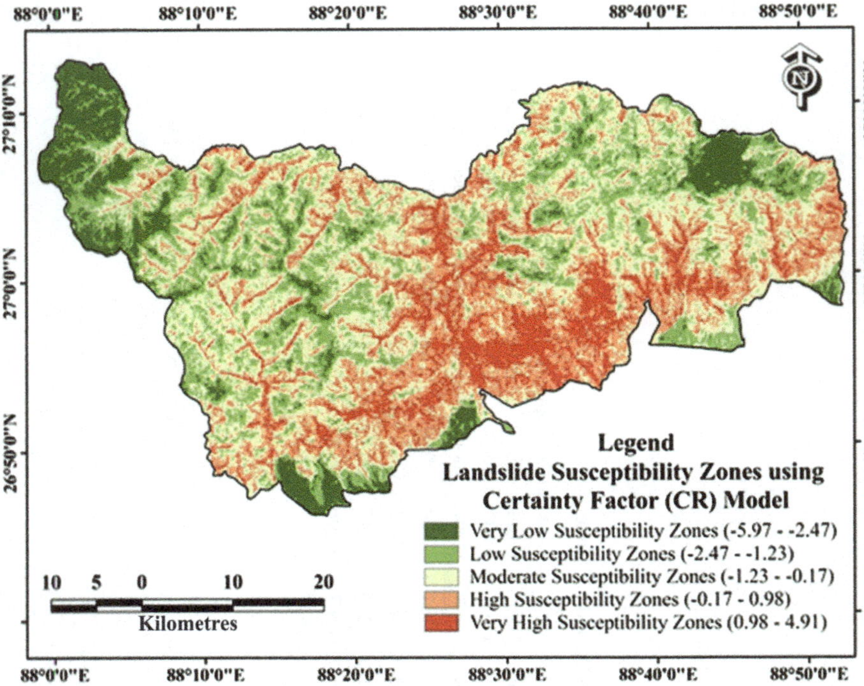

Fig. 6.3 Landslide susceptibility zonation map using certainty factor approach

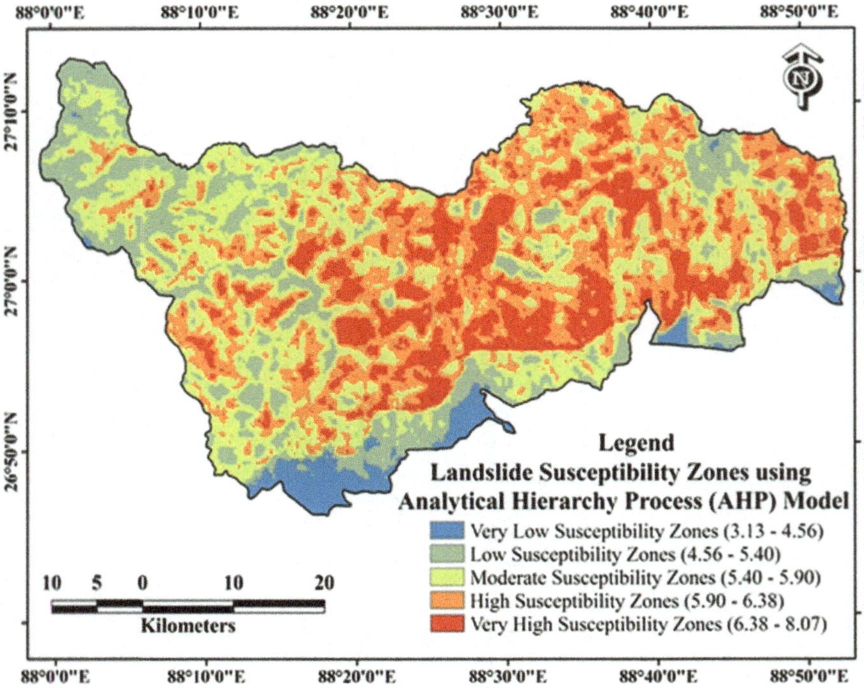

Fig. 6.4 Landslide susceptibility zonation map using analytical hierarchy process (AHP)

6.3.4 *Comparison Between Weighted Overlay Analysis (WOA), Certainty Factor (CF), and Analytical Hierarchy Process (AHP) Models*

The prepared landslide susceptibility map of Darjeeling Himalaya using weighted overlay index model, certainty factor approach, and analytical hierarchy processes was classified into very low, low, moderate, high, and very high landslide susceptibility zones. All the approaches showed a positive relationship between landslide susceptibility and probability of landslides. The model validation using ROC curve revealed that weighted index overlay model, certainty factor approach, and analytical hierarchy process having the accuracy level of 65.4%, 81.2%, and 67.5%, respectively (Fig. 6.5). Certainty factor approach showed higher accuracy level, and this model could be adopted for the mapping and modelling of landslide susceptibility of Darjeeling Himalaya. The frequency ratio plot of all the models introduced a positive relationship between landslide susceptibility and frequency ratio as well as landslide probability (Fig. 6.6). In certainty factor approach, a strong positive relationship was observed between high and very high landslide susceptibility and frequency ratio in comparison to other two models, i.e. weighted index overlay and analytical hierarchy process.

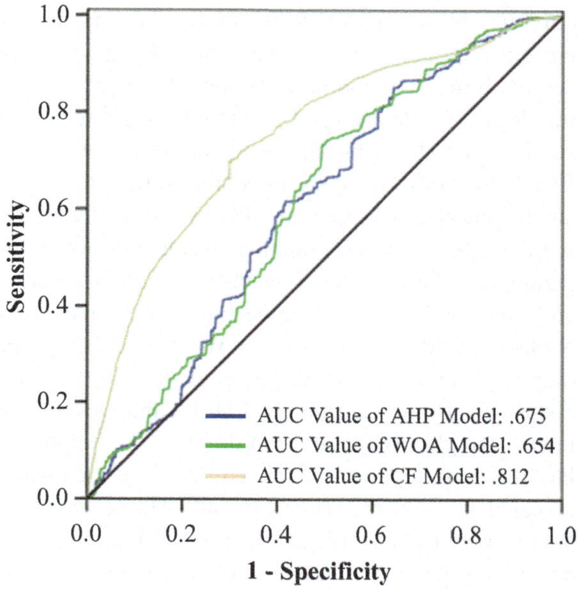

Fig. 6.5 ROC curve and validation result of WOA model, CF model, and AHP model

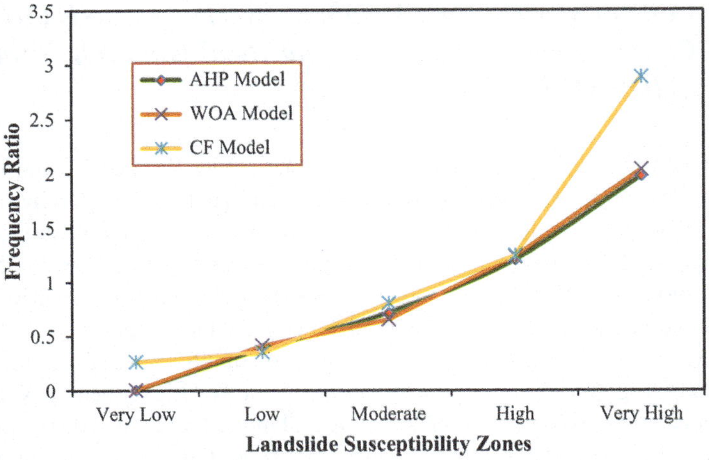

Fig. 6.6 Frequency ratio plot of WOA model, CF model, and AHP model

6.4 Conclusion

Darjeeling Himalaya exhibits a wide range of variation in terms of geomorphic and geohydrologic parameters. Such variation of geomorphic and geohydrolgic attributes played a significant role in the variation of landslide susceptibility and landslide probability in Darjeeling Himalaya. High to very high landslide susceptibility are common at the places of steep slope, rugged terrain, weak lithology, high intensity of lineaments, drainage density, high intensity of rainfall, south facing slope, high positive and negative curvature, and human intervention. Weighted overlay analysis model introduced initially to integrate various data layers and to generate an output on GIS environment which became a significant approach in the study of any geographical events concerned with large number of attributes/ parameters. This model showed the accuracy level of 70.1% where theme weight and class weight were being assigned based on the relative significance of the factor or class in landslides. To perform WOA model and to prepare landslide susceptibility map of Darjeeling Himalaya, geology, slope, soil, curvature, drainage density, distance to drainage, rainfall, LULC, and elevation were given more priority and theme weight assigned accordingly. The application of AHP and developed landslides susceptibility depicted 72.9% of area was under ROC curve which also revealed a reasonable accuracy result. The accuracy status of AHP is quite high than WOA model. In AHP, the hierarchical arrangement of factors were made by developing pair-wise comparison matrix and where landslide conditioning factors of geology, rainfall, drainage density, distance to drainage, slope, elevation, soil, and curvature were ranked as first, second, third, fourth, fifth, sixth, seventh, and eighth based on derived eigen vector. The developed couple-comparing matrix showed the consistency ratio of 0.016 which is accepted. Certainty factor approach and prepared landslide susceptibility map of Darjeeling Himalaya depicted accuracy level of 81.2% which is greater than WOA model and AHP model. It could be assumed that certainty factor approach and associated landslide susceptibility assessment and prediction of Darjeeling Himalaya were being recognized as most reliable and scientific in comparison to WOA model AHP model and CF model could be adopted to identify suitable sites for ensuring developmental activities.

References

Abramson, L. Y., Alloy, L. B., & Hogan, M. E. (1995). Cognitive/personality subtypes of depression: Theories in search of disorders. *Cognitive Therapy and Research, 21*, 247–265.

Ayalew, L., Yamagishi, H., Marui, H., & Kanno, T. (2005). Landslides in Sado island of Japan: Part II, GIS-based susceptibility mapping with comparisons of results from two methods and verifications. *Engineering Geology, 81*, 432–445.

Basu, S. R., & Ghatowar, L. (1988). Landslide in the Lish Basin of the Eastern Himalayas and their control. In *Geomorphology and environment*. Allahabad: The Allahabad Geographical Society.

Biaghi, E., Luzi, L., Mandella, P., Pergalani, F., & Rampini, A. (1998). Slope instability zonation: A comparison between certainty factor and fuzzydempster shafer approaches. *Natural Hazars, 17*, 77–97.

Chau, K. T., Sze, Y. L., Fung, M. K., Wong, W. Y., Fong, E. L., & Chan, L. C. P. (2004). Landslide hazard analysis for Hong Kong using landslide inventory and GIS. *Computers & Geosciences, 30*, 429–443.

Chung, C. F., & Fabbri, A. G. (1993). The representation of geoscience information for data integration. *Nonrenewable Resources, 2*(2), 122–139.

Dai, F. C., & Lee, C. F. (2002). Landslide characteristics and slope instability modelling using GIS, Lantau Island, Hong Kong. *Geomorphology, 42*, 213–228.

Donati, L., & Turrini, M. C. (2002). An objective method to rank, the importance of the factors predisposing to landslides with the GIS methodology: Application to an area of the Apennines (Valnerina; Perugia, Italy). *Engineering Geology, 63*, 277–289.

Dou, J., Oguchi, T., Hayakawa, Y. S., Uchiyama, S., Saito, H., & Paudel, U. (2014). GIS-based landslide susceptibility mapping using a certainty factor model and its validation in the Chuetsu area, central Japan. In *Landslide science for a safer geoenvironment* (pp. 419–424). New York: Springer.

Guzzetti, F., Carrara, A., Cardinali, M., & Reichenbach, P. (1999). Landslide hazard evaluation: A review of current techniques and their application in a multi-scale study, Central Italy. *Journal of Geomorphology, 31*, 181–216.

Heckeman. (1986). Probabilistic interpretation of MYCIN's certainty factors. In L. N. Kanal & J. F. Lemmer (Eds.), *Uncertainty in artificial intelligence B* (pp. 298–311). New York: Elsevier.

Ilia, I., Koumantakis, I., Rozos, D., Koukis, G., & Tsangaratos, P. (2015). A geographical information system (GIS) based probabilistic certainty factor approach in assessing landslide susceptibility: The case study of Kimi, Euboea, Greece. In *Engineering geology for society and territory* (Vol. 2, pp. 1199–1204). Berlin: Springer.

Jibson, R. W., Harp, E. L., & Michael, J. A. (2000). A method for producing digital probabilistic seismic landslide hazard maps. *Engineering Geology, 58*, 271–289.

Kanungo, D. P., Sarkar, S., & Sharma, S. (2011). Combining neural network with fuzzy, certainty factor and likelihood ratio concepts for spatial prediction of landslides. *Natural Hazards, 59*(3), 1491–1512.

Kienholz, H. (1978). Maps of geomorphology and natural hazards of Grindelwald, Switzerland: Scale 1:10,000. *Arctic and Alpine Research, 10*, 169–184.

Komac, M. (2006). A landslide susceptibility model using the Analytical Hierarchy Process method and multivariate statistics in peialpine Slovenia. *Geomorphology, 74*, 17–28.

Liu, M., Chen, X., & Yang, S. (2014). Collapse landslide and mudslide hazard zonation. In *Landslide science for a safer geoenvironment* (pp. 457–462). Berlin: Springer.

Malczewski, J. (1999). *GIS and multi-criteria decision analysis* (p. 392). New York: Wiley.

Mandal, S., & Maiti, R. (2011). Landslide susceptibility analysis of Shivkhola Watershed, Darjeeling: A remote sensing & GIS Based Analytical Hierarchy Process (AHP). *Journal of Indian Society of Remote Sensing*. https://doi.org/10.10007/s12524-011-0160-9.

Moradi, M., Bazyar, M. H., & Mohammadi, Z. (2012). GIS-based landslide susceptibility mapping by AHP method: A case study, Dena City, Iran. *Journal of Basic and Applied Scientific Research, 2*, 6715–6723.

Pourghasemi, H. R., Pradhan, B., Gokceoglu, C., Mohammadi, M., & Moradi, H. R. (2013). Application of weights-of-evidence and certainty factor models and their comparison in landslide susceptibility mapping at Haraz Watershed, Iran. *Arabian Journal of Geosciences, 6*(7), 2351–2365.

Praise, M., & Jibson, R. W. (2000). A seismic landslide susceptibility rating of geologic units based on analysis of characteristics of landslide triggered by the January 17, 1994 Northridge, California, earthquake. *Engineering Geology, 58*, 251–270.

Rautelal, P., & Lakheraza, R. C. (2000). Landslide risk analysis between Giri and Tons Rivers in Himachal Himalaya, India. *International Journal of Applied Earth Observation and Geoinformation, 2*, 153–160.

Saaty, T. L. (1977). A scaling method for priorities in hierarchical structures. *Journal of Mathematical Psychology, 15*, 234–281.

Saaty, T. L. (1980). *The analytical hierarchy process*. New York: McGraw-Hill.

Saaty, T. L. (1994). *Fundamentals of decision making and priority theory with analytic hierarchy process*. Pittsburgh: RWS Publications.

Saaty, T. L., & Vargas, G. L. (1991). *Prediction, projection and forecasting*. Dordrecht: Kluwer Academic.

Saaty, T. L., & Vargas, G. L. (2001). *Models, methods, concepts and applications of the analytic hierarchy process*. Dordrecht: Kluwer.

Shortliffe, E. H., & Buchanan, B. G. (1975). A model of inexact reasoning in medicine. *Mathematical Biosciences, 23*(3), 351–379.

Spiker, E. C., & Gori, P. L. (2000). *National landslide hazards mitigation strategy: A framework for loss reduction* (p. 59). Reston: Department of the Interior, U.S. Geological Survey.

Sujatha, E. R., Rajamanickam, G. V., & Kumaravel, P. (2012). Landslide susceptibility analysis using probabilistic certainty factor approach: A case study on Tevankarai stream watershed, India. *Journal of Earth System Sciences,* ***121*****(5)**, 1337–1350.

Varnes, D. J. (1978). Slope move types and processes. In R. L. Schuster & R. J. Krized (Eds.), *Landslide analysis and control* (pp. 12–33). New York: National Academy of Science.

Varnes, D. J. (1981). Slope stability problems of the circum Pacific region as related to mineral and energy resource. In M. T. Halbouty (Ed.), *Energy resources of the Pacific region. American Association of Petroleum Geologists Studies in Geology. No. 12* (pp. 489–505). Tulsa, Okla: American Association of Petroleum Geologist.

Yalcin, A. (2008). GIS-based landslide susceptibility mapping using analytical hierarchyprocess and bivariate statistics in Ardesen (Turkey): Comparisons of results andconfirmations. *Catena, 72*, 1–12.

Zhou, C. H., Lee, C. F., Li, J., & Xu, Z. W. (2002). On the spatial relationship between landslides and causative factors on Lantau Island, Hong Kong. *Geomorphology, 43*, 197–207.

Chapter 7
Knowledge-Driven Statistical Approach for Landslide Susceptibility Assessment Using GIS and Fuzzy Logic (FL) Approach

7.1 Introduction

Since second half of twentieth century, Darjeeling Himalaya exhibited a wide range of human interferences such as expansion of human settlement, construction of roads, expansion of tea garden areas, expansion of tourism industry, and large-scale slope clearances have aggravated the problems of landslides at different scales. The development of roads and slope modification is a common phenomenon in Darjeeling Himalaya. Small-scale landslide scars are quite common with high frequency in the mountain environment (Dai et al. 2001). Landslide susceptibility to slope failure or slope instability is basically caused due to complex geological setting combined with various geomorphological, hydrological, and geo-technical parameters. Landslide susceptibility mapping technique is associated with various factors which cause a suitable environment to move the slope materials downward. All factors concerned with slope materials movement is regarded as predisposing factors on the basis of which landslide susceptibility mapping is being made on GIS environment. Carrara et al. (1991) and Dai and Lee (2002) suggested that the predisposing factors are intrinsic in character and characteristic of land and soil. Slope, relief, soil, geology, land use and land cover, and drainage network are the common predisposing factors. Predisposing factors can be regarded as "causative factors", "casual factors", "intrinsic factors", "conditioning factors", "quasi-static factors", and "preparatory factors" which were propounded by Varnes and IAEG (1984), Zhou et al. (2002), Zezere et al. (1999), Atkinson and Massari (1998), and Dai and Lee (2002). There are many approaches to assess slope stability and landslide hazards (Sidle 1991; Montgomery and Dietrich 1989, 1994; Carrara et al. 1991; Dietrich et al. 1993). Montgomery and Dietrich (1994) combine a contour-based steady-state hydrologic model with the infinite slope stability model (simplified for cohesion less soils) to define slope

S. Mandal, S. Mondal, *Statistical Approaches for Landslide Susceptibility Assessment and Prediction*, https://doi.org/10.1007/978-3-319-93897-4_7

stability classes based upon slope and specific catchment area. In the present day, various statistical techniques and methods were used to assess landslide susceptibility using landslide causative factors and historical landslide data. But, there are few shortcomings in using statistical models and preparing landslide susceptibility map with reasonable accuracy. Statistical models are associated with training data corresponding to landslide occurrences in the concerned region. The preparation of training data set and its quality depends upon the nature of field investigation adopted by researchers. In most of the cases, proper field investigation with necessary accessories is not being incorporated and where the shortcoming of the statistical model exists. Basically, the data-oriented landslide models are extremely data hungry (Mitchell 1997). To obtain historical landslide statistics and landslide inventory map of the concerned study area intensive field investigation with GPS, Google earth images, satellite imageries, and topographical maps are very much essential. Statistical models are based on linear models which represents the relationship between variables (Van Westen et al. 2003). The relationship is not always linear and it sometimes becomes non-linear (Liu et al. 2004; Zhu et al. 2004). All the statistical models must consider expert knowledge for data retrieving and processing which can help prepare landslide susceptibility map with good accuracy result. The expert knowledge-based approach considers the relationship between predisposing factors and landslide susceptibility map from local landslide experts to find out the relationship (Zhu et al. 2004). The expert knowledge-based approach is dealt with the complicated of non-linear relationship between predisposing factors and landslide susceptibility. This relationship is extracted under fuzzy logic which represents as a set of fuzzy membership functions. In fuzzy membership functions, the relationship between landslides susceptibility and predisposing factors are being analysed. The predisposing factors were assessed with the help of GIS and the compilations of all the factors in a raster domain were made also on GIS environment (Remondo and Oguchi 2009). Sezer et al. (2011), Pourghasemi et al. (2012), Pradhan et al. (2009), Pradhan (2010, 2011), Kanungo et al. (2006), and Bui et al. (2012) applied fuzzy logic approach in landslide susceptibility studies using RS and GIS.

To develop landslide susceptibility model, various statistical, analytical, and numerical approaches have been applied by researchers (Ermini et al. 2005; Tofani et al. 2006; Ren et al. 2011; Lepore et al. 2012). Amongst all these approaches bivariate statistical approach, analytical hierarchy process, frequency ratio model, certainty factor approach, weight of evidence model, logistic regression model, artificial neural network model, fuzzy logic, and support vector machine are widely accepted in modelling landslide susceptibility map using RS and GIS. In the present study, fuzzy logic has been applied to prepare landslide susceptibility zonation map of Darjeeling Himalaya where landslide predisposing factors were being presented by a value ranges between 0 and 1 (Champatiray 2000 and Champatiray et al. 2007). Then, the values were integrated using fuzzy gamma operator or fuzzy algebraic sum to produce landslide susceptibility map.

7.2 Materials and Methods

The preparation of the landslides susceptibility map of Darjeeling Himalaya is associated with the creation of data layers, estimation of fuzzy membership value using frequency ratio method and cosine amplitude method, application of gamma operator on GIS environment, and validation of the result using ROC curve (Fig. 7.1). All the data layers, i.e. elevation, slope aspect, slope angle, slope curvature, geology, soil, lineament density, distance to lineament, drainage density, distance to drainage, stream power index (SPI), topographic wetted index (TWI), rainfall, normalized differential vegetation index (NDVI) and land use and land cover (LULC) were made in consultation with topographical map, Google earth imageries, satellite imageries, SRTM DEM and some other authorized maps. The detailed description of the methodology for generating data layers are stated in Chap. 2.

7.2.1 *Estimation of Fuzzy Membership Value*

Membership value was determined with the help of membership function. Fuzzy membership function is a function that maps all given elements to an interval values between <0 and 1> (Eq. 7.1).

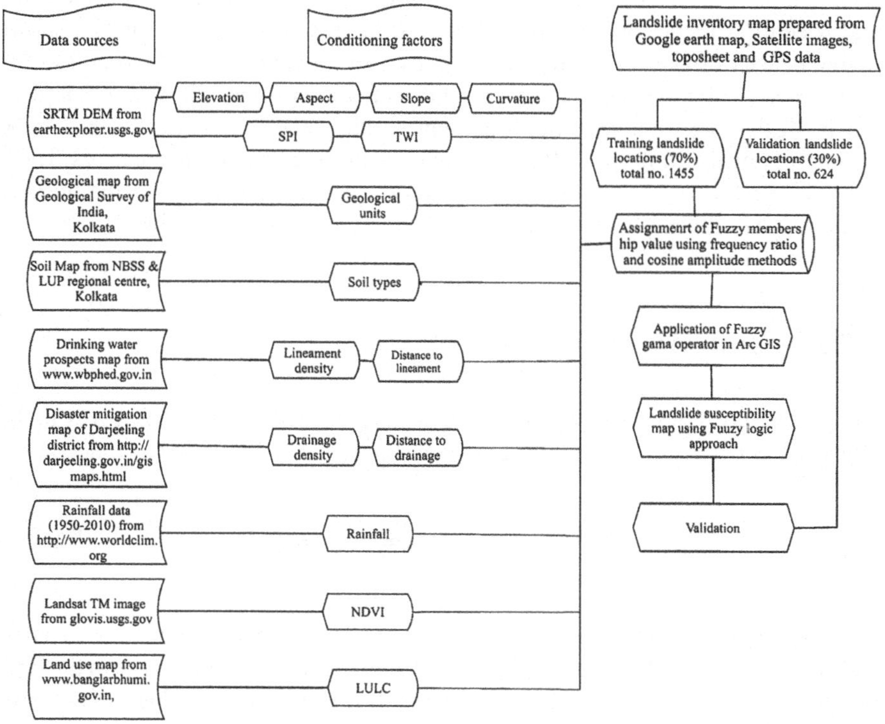

Fig. 7.1 Flow chart of the methodology to prepare landslide susceptibility map

$$\mu_A : U \rightarrow 0,1 \tag{Eq. 7.1}$$

where μ_A is a membership function, U is a set of elements. Then for each $x \epsilon U$, $\mu_A(x)$ is membership value of the element x to the set A (Zadeh 1965). Two functions were being used for computing the fuzzy membership values:

1. Frequency ratio
2. Cosine amplitude

7.2.2 *Frequency Ratio*

Frequency ratio provides proportion of landslide cells in the specific category for each input layers. It can be described as ratio of relative frequency of landslide cells in a category (an attribute class) to the relative frequency of all landslide cells in the area (Eq. 7.2):

$$\text{FR} = \frac{N\ \text{cell}(\text{Li}) / N\ \text{cell}(\text{ci})}{N\ \text{cell}(\text{L}) / N\ \text{cell}(\text{C})} \tag{Eq. 7.2}$$

where N cell (Li) is the number of landslide cells in the category i, N cell (Ci) is the total number of cells in the category i, N cell (L) is total number of landslide cells, and N cell (C) is the total number of cells. If the result is higher than 1, it shows higher density of landslide cells in the category than overall in the data set. Results lower than 1 point to categories that have density of landslide cells lower than density in the data set. To transform FR to membership values, those outputs have to be normalized by dividing each FR by maximal FR in the given group of classes. Then, the membership values are created between <0 and 1> where higher the number greater is the influence of this category on landslide occurrence (Table 7.1).

7.2.3 *Cosine Amplitude*

Another method for determining the fuzzy membership values of each category of the input data layers is cosine amplitude method (Eq. 7.3) which is stated as follows:

$$\text{CA} = \frac{N\ \text{cell}(\text{Li})}{\sqrt{N\ \text{cell}(\text{Ci}) \cdot N\ \text{cell}(\text{L})}} \tag{Eq. 7.3}$$

In this case, the membership value is calculated as ratio between number of landslide cells in the data layers category and the square root of its product with the total number of landslide pixels in the data set. Unlike FR, the output values do not have to be normalized because they already fall between <0 and 1> (Table 7.1).

Table 7.1 Fuzzy membership values using frequency ratio (FR) and cosine amplitude (CA) methods

Conditioning factor	Class	Number of pixel	Total pixel (%)	Number of landslide occurrence pixels	Landslide occurrence pixels (%)	Frequency ratio value	Fuzzy membership value using frequency ratio (FR) method	Fuzzy membership value using cosine amplitude (CA) method
Elevation (m)	130–470	409,786	15.89	465	9.53	0.60	0.33	0.31
	470–770	437,151	16.96	1519	31.11	1.83	1.00	0.56
	770–1100	437,588	16.97	1259	25.78	1.52	0.83	0.51
	1100–1300	246,682	9.57	445	9.11	0.95	0.52	0.30
	1300–1600	342,538	13.29	465	9.53	0.72	0.39	0.31
	1600–1900	260,070	10.09	178	3.65	0.36	0.20	0.19
	1900–2200	171,257	6.64	268	5.50	0.83	0.45	0.23
	2200–2600	142,581	5.53	207	4.25	0.77	0.42	0.21
	2600–3000	80,867	3.14	69	1.42	0.45	0.25	0.12
	3000–3600	49,605	1.92	6	0.12	0.06	0.03	0.04
Aspect	Flat (−1)	231	0.01	0	0.00	0.00	0.00	0.00
	North (0–22.5)	136,124	5.28	90	1.84	0.35	0.23	0.14
	Northeast (22.5–67.5)	296,730	11.51	706	14.47	1.26	0.85	0.38
	East (67.5–112.5)	3959,73	15.36	890	18.22	1.19	0.80	0.43
	Southeast (112.5–157.5)	413,694	16.05	894	18.30	1.14	0.77	0.43
	South (157.5–202.5)	386,071	14.97	777	15.91	1.06	0.71	0.40
	Southwest (202.5–247.5)	299,310	11.61	844	17.29	1.49	1.00	0.42
	West (247.5–292.5)	266,605	10.34	396	8.11	0.78	0.52	0.28
	Northwest (292.5–337.5)	250,427	9.71	252	5.16	0.53	0.36	0.23
	North (337.5–360)	132,961	5.16	34	0.70	0.14	0.09	0.08

(continued)

Table 7.1 (continued)

Conditioning factor	Class	Number of pixel	Total pixel (%)	Number of landslide occurrence pixels	Landslide occurrence pixels (%)	Frequency ratio value	Fuzzy membership value using frequency ratio (FR) method	Fuzzy membership value using cosine amplitude (CA) method
Slope (degree)	0–4.339	187,019	7.25	134	2.74	0.38	0.15	0.17
	4.339–8.677	255,476	9.91	403	8.26	0.83	0.34	0.29
	8.677–12.515	343,632	13.33	590	12.08	0.91	0.37	0.35
	12.515–15.853	374,691	14.53	668	13.69	0.94	0.38	0.37
	15.853–19.024	427,124	16.57	746	15.27	0.92	0.37	0.39
	19.024–22.027	382,230	14.83	704	14.42	0.97	0.39	0.38
	22.027–25.031	278,609	10.81	573	11.73	1.09	0.44	0.34
	25.031–28.202	189,918	7.37	459	9.39	1.28	0.52	0.31
	28.202–32.373	107,393	4.17	501	10.27	2.47	1.00	0.32
	32.373–42.553	32,035	1.24	104	2.14	1.72	0.70	0.15
Curvature	−0.644 to −0.238	48,330	1.87	159	3.26	1.74	0.73	0.18
	−0.238 to −0.155	168,473	6.53	765	15.66	2.40	1.00	0.40
	−0.155 to −0.095	286,017	11.09	855	17.51	1.58	0.66	0.42
	−0.095 to −0.040	390,224	15.14	859	17.59	1.16	0.48	0.42
	−0.040 to 0.015	577,690	22.41	672	13.77	0.61	0.25	0.37
	0.015–0.064	406,708	15.78	548	11.23	0.71	0.30	0.34
	0.064–0.120	330,604	12.82	435	8.91	0.69	0.29	0.30
	0.120–0.185	234,280	9.09	330	6.75	0.74	0.31	0.26
	0.185–0.267	108,702	4.22	192	3.94	0.93	0.39	0.20
	0.267–0.756	27,099	1.05	67	1.38	1.31	0.55	0.12

Geology	Alluvium	104,604	4.06	27	0.55	0.14	0.08	0.07
	Baxa Series-Slates, Schists, Dolomite, Quartzites	2344	0.09	4	0.08	0.84	0.49	0.03
	Daling Series-Slates, Schists, Quartzites	712,203	27.62	1975	40.46	1.46	0.85	0.64
	Damuda (Gondwana)	35,988	1.40	116	2.38	1.71	1.00	0.15
	Darjeeling Gneiss	1,523,340	59.09	2631	53.90	0.91	0.53	0.73
	Nahan Group (Tertiary) Limestone	199,648	7.74	128	2.63	0.34	0.20	0.16
Soil	Tista River	14,861	0.58	0	0.00	0.00	0.00	0.00
	W001	261,470	10.14	492	10.08	0.99	0.42	0.32
	W002	1,434,692	55.65	2183	44.71	0.80	0.34	0.67
	W003	221,855	8.61	183	3.75	0.44	0.19	0.19
	W004	446,612	17.32	1967	40.30	2.33	1.00	0.63
	W006	75,842	2.94	1	0.02	0.01	0.00	0.01
	W007	85,141	3.30	17	0.34	0.10	0.04	0.06
	W008	35,338	1.37	39	0.80	0.58	0.25	0.09
	W009	2314	0.09	0	0.00	0.00	0.00	0.00
Drainage density (km/sq. km)	0–0.694	243,293	9.44	266	5.44	0.58	0.45	0.23
	0.694–1.210	409,926	15.90	732	14.99	0.94	0.73	0.39
	1.210–1.620	469,031	18.19	879	18.00	0.99	0.77	0.42
	1.620–2.010	498,755	19.35	1202	24.62	1.27	0.98	0.50
	2.010–2.370	393,587	15.27	753	15.42	1.01	0.78	0.39
	2.370–2.730	272,462	10.57	665	13.62	1.29	1.00	0.37
	2.730–3.140	166,766	6.47	279	5.72	0.88	0.68	0.24
	3.140–3.680	76,489	2.97	83	1.69	0.57	0.44	0.13
	3.680–4.450	34,741	1.35	20	0.41	0.31	0.24	0.06
	4.450–6.560	13,076	0.51	4	0.08	0.16	0.12	0.03

(continued)

Table 7.1 (continued)

Conditioning factor	Class	Number of pixel	Total pixel (%)	Number of landslide occurrence pixels	Landslide occurrence pixels (%)	Frequency ratio value	Fuzzy membership value using frequency ratio (FR) method	Fuzzy membership value using cosine amplitude (CA) method
Lineament density (km/sq. km)	0–0.056	1,309,367	50.79	2325	47.63	0.94	0.65	0.69
	0. 056–0.168	300,996	11.67	482	9.87	0.85	0.59	0.31
	0.168–0.287	230,519	8.94	577	11.81	1.32	0.91	0.34
	0.287–0.399	192,603	7.47	337	6.90	0.92	0.63	0.26
	0.399–0.505	162,361	6.30	342	7.00	1.11	0.77	0.26
	0.505–0.610	141,605	5.49	390	7.98	1.45	1.00	0.28
	0.610–0.716	102,252	3.97	248	5.09	1.28	0.88	0.23
	0.716–0.835	73,150	2.84	118	2.42	0.85	0.59	0.16
	0.835–0.966	48,815	1.89	53	1.08	0.57	0.39	0.10
	0.966–1.590	16,458	0.64	11	0.22	0.35	0.24	0.05
Distance from lineament (m)	0–20	23,691	0.92	42	0.86	0.93	0.74	0.09
	20–50	36,643	1.42	61	1.25	0.88	0.70	0.11
	50–100	63,438	2.46	119	2.43	0.99	0.79	0.16
	100–400	406,582	15.77	823	16.86	1.07	0.86	0.41
	400–1000	787,207	30.53	1426	29.22	0.96	0.77	0.54
	1000–2000	781,676	30.32	1448	29.66	0.98	0.78	0.54
	2000–3500	369,169	14.32	872	17.87	1.25	1.00	0.42
	3500–5000	95,146	3.69	91	1.87	0.51	0.41	0.14
	5000–6500	10,670	0.41	0	0.00	0.00	0.00	0.00
	6500–8500	3904	0.15	0	0.00	0.00	0.00	0.00

Distance to drainage (m)	0–10	92,856	3.60	210	4.31	1.20	1.00	0.21
	10–30	184,399	7.15	412	8.43	1.18	0.98	0.29
	30–70	356,416	13.82	754	15.44	1.12	0.93	0.39
	70–150	628,426	24.38	1291	26.45	1.09	0.91	0.51
	150–300	763,840	29.63	1599	32.76	1.11	0.93	0.57
	300–700	498,808	19.35	600	12.29	0.64	0.53	0.35
	700–1200	45,871	1.78	11	0.23	0.13	0.11	0.05
	1200–1700	5659	0.22	4	0.08	0.38	0.32	0.03
	1700–2300	1593	0.06	0	0.00	0.00	0.00	0.00
	2300–3000	259	0.01	0	0.00	0.00	0.00	0.00
SPI	−5.939 to −3.140	14,266	0.55	20	0.40	0.72	0.35	0.06
	−3.140 to −2.496	57,082	2.21	92	1.88	0.85	0.41	0.14
	−2.496 to −2.013	132,095	5.12	221	4.53	0.88	0.43	0.21
	−2.013 to −1.595	223,338	8.66	379	7.76	0.90	0.43	0.28
	−1.595 to −1.209	296,386	11.50	482	9.87	0.86	0.42	0.31
	−1.209 to −0.791	361,250	14.01	612	12.55	0.90	0.43	0.35
	−0.791 to −0.373	379,786	14.73	691	14.16	0.96	0.46	0.38
	−0.373 to 0.014	571,397	22.16	1040	21.29	0.96	0.46	0.46
	0.014–0.400	391,010	15.17	752	15.41	1.02	0.49	0.39
	0.400–2.266	151,516	5.88	593	12.15	2.07	1.00	0.35

(continued)

Table 7.1 (continued)

Conditioning factor	Class	Number of pixel	Total pixel (%)	Number of landslide occurrence pixels	Landslide occurrence pixels (%)	Frequency ratio value	Fuzzy membership value using frequency ratio (FR) method	Fuzzy membership value using cosine amplitude (CA) method
TWI	−6907.755	231	0.01	0	0.00	0.00	0.00	0.00
	−6907.755 to −2849.717	848	0.03	0	0.00	0.00	0.00	0.00
	−2849.717 to −1226.502	2601	0.10	0	0.00	0.00	0.00	0.00
	−1226.502 to −520.757	6953	0.27	0	0.01	0.04	0.04	0.00
	−520.757 to −167.884	36,026	1.40	15	0.32	0.23	0.21	0.06
	−167.884 to −26.735	454,856	17.64	639	13.09	0.74	0.67	0.36
	−26.735 to 43.840	1,987,528	77.09	4150	85.00	1.10	1.00	0.92
	43.840–255.564	76,829	2.98	67	1.37	0.46	0.42	0.12
	255.564–714.298	10,964	0.43	10	0.21	0.49	0.45	0.05
	714.298–2090.502	1289	0.05	0	0.00	0.00	0.00	0.00
Mean annual rainfall (mm)	1076–1478	54,493	2.11	40	0.82	0.39	0.23	0.09
	1478–1798	70,178	2.72	43	0.89	0.33	0.19	0.09
	1798–2097	73,731	2.86	82	1.69	0.59	0.35	0.13
	2097–2355	88,710	3.44	287	5.89	1.71	1.00	0.24
	2355–2551	268,934	10.43	199	4.07	0.39	0.23	0.20
	2551–2705	551,234	21.38	909	18.63	0.87	0.51	0.43
	2705–2860	608,233	23.59	1341	27.48	1.16	0.68	0.52
	2860–3056	423,558	16.43	1165	23.87	1.45	0.85	0.49
	3056–3293	277,977	10.78	628	12.86	1.19	0.70	0.36
	3293–3705	161,077	6.25	187	3.82	0.61	0.36	0.20

NDVI	−0.200–0.037	34,132	1.32	324	6.64	5.03	1.00	0.26
	0.037–0.110	77,506	3.01	482	9.87	3.28	0.65	0.31
	0.110–0.166	167,231	6.49	520	10.65	1.64	0.33	0.33
	0.166–0.214	273,220	10.60	683	13.98	1.32	0.26	0.37
	0.214–0.263	381,618	14.80	759	15.54	1.05	0.21	0.39
	0.263–0.315	451,331	17.51	753	15.43	0.88	0.17	0.39
	0.315–0.368	428,060	16.60	587	12.02	0.72	0.14	0.35
	0.368–0.427	364,538	14.14	425	8.70	0.62	0.12	0.30
	0.427–0.493	257,616	9.99	245	5.02	0.50	0.10	0.22
	0.493–0.667	142,875	5.54	105	2.15	0.39	0.08	0.15
LULC	Social Forestry	309,800	12.02	453	9.27	0.77	0.10	0.30
	Agricultural Single Crop	331,856	12.87	537	10.99	0.85	0.11	0.33
	Forest	1,430,826	55.50	2978	61.00	1.10	0.15	0.77
	Rural Settlement	126,193	4.89	150	3.07	0.63	0.08	0.17
	Urban Settlement	16,672	0.62	18	0.36	0.58	0.08	0.06
	Agricultural Plantation (Tea)	337,588	13.09	514	10.53	0.80	0.11	0.32
	Wasteland Dry/Barren Land	399	0.02	7	0.15	7.50	1.00	0.04
	Public Utility and Facility	3052	0.12	0	0.00	0.00	0.00	0.00
	Wastelands with Scrub	21,740	0.84	225	4.61	5.49	0.73	0.21

7.2.4 *Fuzzy Operators*

Several fuzzy operators exist for combining membership functions. Best-known operators are AND and OR, but both of them suffer with problem that one of combined sets have significant impact on result of such combination while the other sets do not have such influence. In case of operator AND, minimum of all values is the one that defines output and in case of OR operator it is the maximum value. Fuzzy algebraic product tends to produce output function lower or equal to the lowest function given, while fuzzy algebraic sum is complementary to the former, so it provides output function higher than all the inputs but never higher than 1. In the present study, gamma operator was applied for combining membership functions.

Fuzzy gamma operation is defined by:

$$\mu_{\gamma} = \left(\mu_{\text{sum}}\right) k^{\gamma} . \left(\mu_{\text{product}}\right).^{1-\gamma} \quad \text{(Eq. 7.4)}$$

The exponent γ, which is a number from <0, 1> interval, allows optimization of the membership combination. Setting it to the extremes of the interval give either fuzzy algebraic sum ($\gamma = 1$) or fuzzy algebraic product ($\gamma = 0$).

7.2.4.1 Application of Fuzzy Gamma Operators

To perform fuzzy gamma operation, several gamma operator (k) values, i.e. 0.50, 0.70, 0.80, 0.90, 0.95, and 0.975 were tested. It resulted that the best performance (AUC) was achieved by $\gamma = 0.975$, and made it as a parameter of choice for producing final landslide susceptibility model and landslide susceptibility map. For assigning membership values, frequency ratio could be preferred over cosine amplitude.

7.3 Result and Discussion

7.3.1 *Frequency Ratio and Landslide Susceptibility Zonation of Darjeeling Himalaya*

The application of frequency ratio method for deriving fuzzy membership value and performing gamma operation on GIS platform classified Darjeeling Himalaya in five landslide susceptibility zones, i.e. very low, low, moderate, high, and very high landslide susceptibility (Fig. 7.2). The study revealed that Darjeeling Himalaya is dominated by high landslide susceptibility and followed by moderate, very high, low, and very low (Table 7.2). It was also observed that very high and high landslide susceptibility zones were affected mostly by landslide activities. Frequency ratio value of very high, high, moderate, low, and very low landslide susceptibility zones

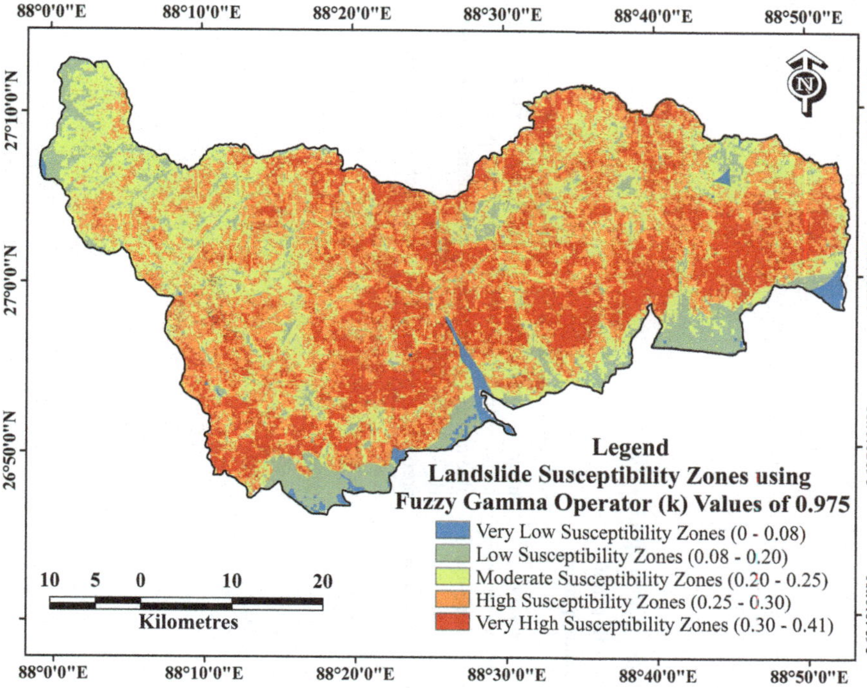

Fig. 7.2 Landslide susceptibility zonation map using fuzzy membership value based on frequency ratio method

are 2.84, 0.84, 0.34, 0.18, and 0.12, respectively. The derived frequency ratio value demonstrated that the landslide probability in the zones of very high, high, and moderate landslide susceptibility is high in comparison to low and very low landslide susceptibility.

7.3.2 Cosine Amplitude Method and Landslide Susceptibility Zonation of Darjeeling Himalaya

Cosine amplitude method-based landslide susceptibility map was classified into five, i.e. very low, low, moderate, high, and very high landslide susceptibility zones (Fig. 7.3). 37.44%, 29.61%, 21.48%, 9.92%, and 1.55% area are registered with high, moderate, very high, low, and very low landslide susceptibility, respectively. In terms of landslide affected area, high landslide susceptibility zone ranks first which was followed by very high, moderate, low, and very low landslide susceptibility (Table 7.2). The estimated frequency ratio of each landslide susceptibility zones depicted that very high and high landslide susceptibility zones having greater probability of landslide phenomena.

Table 7.2 Area and frequency ratio value of different susceptibility classes using fuzzy gamma operator (0.975)

Landslide susceptibility zones	Total pixels using Fuzzy Gamma Operator (0.975)		% of Total pixels		Landslide pixels using Fuzzy Gamma Operator (0.975)		% of Landslide pixels		Frequency ratio value	
	Fuzzy membership value using FR approach	Fuzzy Membership value using CA approach	Fuzzy membership value using FR approach	Fuzzy membership value using CA approach	Fuzzy membership value using FR approach	Fuzzy Membership value using CA approach	Fuzzy membership value using FR approach	Fuzzy membership value using CA approach	Fuzzy membership value using FR approach	Fuzzy membership value using CA approach
Very low	103,483	40,019	4.01	1.55	24	23	0.49	0.48	0.12	0.31
Low	199,453	255,661	7.74	9.92	68	97	1.40	1.98	0.18	0.20
Moderate	797,080	763,289	30.92	29.61	511	708	10.46	14.51	0.34	0.49
High	972,355	965,247	37.72	37.44	1546	2085	31.66	42.71	0.84	1.14
Very high	505,754	553,909	19.62	21.48	2734	1968	56.00	40.32	2.85	1.88

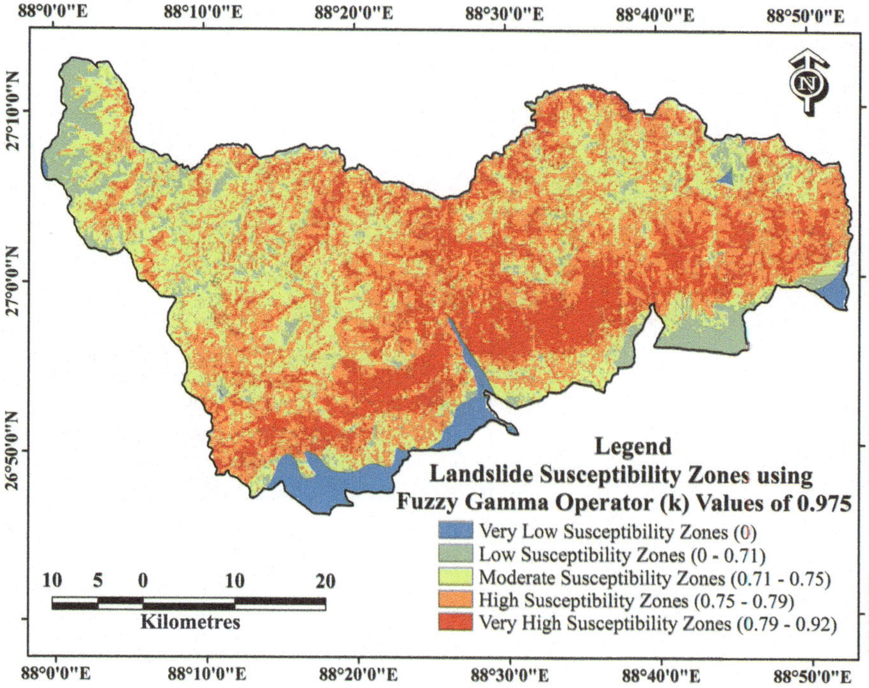

Fig. 7.3 Landslide susceptibility zonation map using fuzzy membership value based on cosine amplitude method

7.3.3 Comparison Between Frequency Ratio-Based Fuzzy and Cosine Amplitude Membership-Based Fuzzy Logic (FL) Approach

Landslide susceptibility map of Darjeeling Himalaya was made based on both frequency ratio-based fuzzy gamma operator and cosine amplitude-based fuzzy gamma operator. Both the maps were being validated with the help of developed ROC curve. Frequency ratio-based fuzzy membership and landslide susceptibility zonation map of Darjeeling Himalaya showed the accuracy level of 80.9% and cosine amplitude-based fuzzy membership and landslide susceptibility depicted the accuracy level of 67.9%. It was found that the frequency ratio-based landslide susceptibility map having high accuracy level than cosine amplitude-based landslide susceptibility (Fig. 7.4).

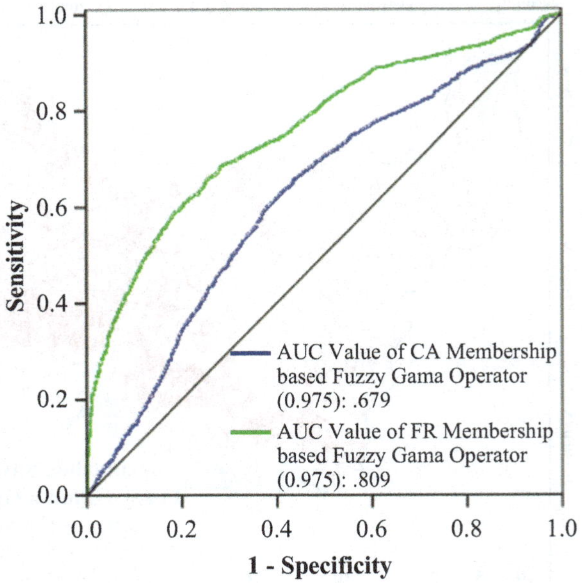

Fig. 7.4 ROC curve and models validation

7.4 Conclusion

The present study is dealt with the application of fuzzy logic and preparation of landslide susceptibility map of Darjeeling Himalaya using the data layers of elevation, slope aspect, slope angle, slope curvature, geology, soil, lineament density, distance to lineament, drainage density, distance to drainage, stream power index (SPI), topographic wetted index (TWI), rainfall, normalized differential vegetation index (NDVI) and land use and land cover (LULC). All the data layers were being integrated based on fuzzy gamma operator and fuzzy membership value on GIS environment. Here, frequency ratio-based estimated fuzzy membership value and derived landslide susceptibility map revealed accuracy level of 80.9%. The fuzzy membership value ranges between <0 and >1. South, southeast, southwest, northeast, and east facing slope having high frequency ratio-based fuzzy membership value as well as greater probability of landslide phenomena. High positive and negative slope curvature showed high fuzzy membership value. There is a positive relationship between slope and fuzzy membership values. The places close to drainage and lineament are characterized by high landslide susceptibility, high fuzzy membership value, and greater probability of landslides. Fuzzy membership values depicted that high drainage density and lineament density areas are very much prone to landslide phenomena and landslide probability. Land use and land cover type of barren land registered fuzzy membership value of 1.

References

Atkinson, P. M., & Massari, R. (1998). Generalized linear modelling of susceptibility to land sliding in the central Apennines, Italy. *Computer and Geosciences, 24*, 373–385.

Bui, D. T., Pradhan, B., Lofman, O., Revhaug, I., & Dick, O. B. (2012). Spatial prediction of landslide hazards in Hoa Binh province (Vietnam): A comparative assessment of the efficacy of evidential belief functions and fuzzy logic models. *Catena, 96*, 28–40.

Carrara, A., Cardinali, M., Detti, R., Guzzetti, F., Pasqui, V., & Reichenbach, P. (1991). GIS techniques and statistical models in evaluating landslide hazard. *Earth Surface Processes and Landforms, 16*, 427–452.

Champatiray, P. (2000). Perationalization of costeffective methodology for landslide hazard zonation using RS and GIS: IIRS initiative. In P. Roy, C. Van Westen, V. Jha, & R. Lakhera (Eds.), *Natural disasters and their mitigation; Remote sensing and geographical information system perspectives* (pp. 95–101). Dehradun, India: Indian Institute of Remote Sensing.

Champatiray, P., Dimri, S., Lakhera, R., & Sati, S. (2007). Fuzzy based methods for landslide hazard assessment in active seismic zone of Himalaya. *Landslides., 4*, 101–110. https://doi.org/10.1007/s1034600600686.

Dai, F. C., & Lee, C. F. (2002). Landslide characteristics and slope instability modelling usingGIS, Lantau Island, Hong Kong. *Geomorphology, 42*, 213–228.

Dai, F. C., Lee, C. F., Li, J., & Xu, Z. W. (2001). Assessment of landslide susceptibility on the natural terrain of Lantau Island, Hong Kong. *Environmental Geology, 40*, 381–391.

Ermini, L., Catani, F., & Casagli, N. (2005). Artificial neural network applied to landslide susceptibility assessment. *Geomorphology, 66*, 327–343.

Kanungo, D. P., Arora, M. K., Sarkar, S., & Gupta, R. P. (2006). A comparative study of conventional, ANN black box, fuzzy and combined neural and fuzzy weighting procedures for landslide susceptibility zonation in Darjeeling Himalayas. *Engineering Geology, 85*, 347–366.

Lepore, C., Kamal, S. A., Shanahan, P., & Bras, R. L. (2012). Rainfall induced landslide susceptibility zonation of Puerto Rico. *Environmental Earth Sciences, 66*, 1–15.

Liu, J. G., Mason, P. J., Clerici, N., Chen, S., Davis, A., Miao, F., Deng, F. L., & Liang, L. (2004). Landslide hazard assessment in the Three Gorges area of the Yangtze River using ASTER imagery: Zigui–Badong. *Geomorphology, 61*, 171–187.

Mitchell, T. M. (1997). *Machine learning* (p. 414). New York: McGraw-Hill.

Montgomery, D. R., & Dietrich, W. E. (1989). Source areas, drainage density and channel initiation. *Water Resources Research, 25*(8), 1907–1918.

Montgomery, D. R., & Dietrich, W. E. (1994). A physically based model for the topographic control on shallow land sliding. *Water Resources Research, 30*(4), 1153–1171.

Pourghasemi, H. R., Pradhan, B., & Gokceoglu, C. (2012). Application of fuzzy logic and analytical hierarchy process (AHP) to landslide susceptibility mapping at Haraz watershed, Iran. *Natural Hazards*. https://doi.org/10.1007/s11069-012-0217-2.

Pradhan, B. (2010). Application of an advanced fuzzy logic model for landslide susceptibility analysis. *International Journal of Computational Intelligence Systems, 3*(3), 370–381.

Pradhan, B. (2011). Manifestation of an advanced fuzzy logic model coupled with geoinformation techniques for landslide susceptibility analysis. *Environmental and Ecological Statistics, 18*(3), 471–493. https://doi.org/10.1007/s10651-010-0147-7.

Pradhan, B., Lee, S., & Buchroithner, M. F. (2009). Use of geospatial data for the development of fuzzy algebraic operators to landslide hazard mapping: A case study in Malaysia. *Applied Geomatics, 1*, 3–15.

Remondo, J., & Oguchi, T. (2009). GIS and SDA applications in geomorphology. *Geomorphology, 111*, 1–3.

Ren, D., Fu, R., Leslie, L. M., & Dickinson, R. E. (2011). Predicting storm-triggered landslides. *Bulletin of the American Meteorological Society, 92*, 129–139.

Sezer, E. A., Pradhan, B., & Gokceoglu, C. (2011). Manifestation of an adaptive neuro-fuzzy model on landslide susceptibility mapping: Klang valley, Malaysia. *Expert Systems with Applications, 38*(7), 8208–8219.

Sidle, R. C. (1991). A conceptual model of changes in root cohesion in response to vegetation management. *Journal of Environmental Quality, 20*(1), 43–52.

Tofani, V., Dapporto, S., Vannocci, P., & Casagli, N. (2006). Infiltration, seepage and slopeinstability mechanisms during the 20–21 November 2000 rainstorm in central Italy Tuscany. *Natural Hazards and Earth System Sciences, 6*, 1025–1033. https://doi.org/10.5194/nhess-6-1025.

Van Westen, C. J., Rengers, N., & Soeters, R. (2003). Use of geomorphological information in indirect landslide susceptibility assessment. *Natural Hazards, 30*, 399–419.

Varnes, D., & IAEG. (1984). *Landslide hazard zonation: A review of principle and practice* (pp. 1–6). Paris: UNESCO.

Zadeh, L. A. (1965). Fuzzy sets. *Information and Control, 8/3*, 338–353.

Zezere, J., Ferreira, A., & Rodrigues, M. (1999). The role of conditioning and triggering factors in the occurrence of landslides: A case study in the area north of Lisbon (Portugal). *Geomorphology, 30*, 133–146.

Zhou, C. H., Lee, C. F., Li, J., & Xu, Z. W. (2002). On the spatial relationship between landslides and causative factors on Lantau Island, Hong Kong. *Geomorphology, 43*, 197–207.

Zhu, A. X., Wang, R. X., Qiao, J., Chen, Y., Cai, Q., & Zhou, C. (2004). Mapping landslide susceptibility in the Three Gorges area, China using GIS, expert knowledge and fuzzy logic (pp. 385–391). In: Y. Chen (Ed.), *GIS and remote sensing in hydrology*. Water Resources and Environment, IAHS Publication 289 (IAHS Red Book), International Association of Hydrological Sciences, Wallingford, UK.

Chapter 8
Comparison Between Statistical Models: A Review and Evaluation

8.1 Introduction

The preparation of landslide susceptibility map using geo-spatial tools for planning and development of mountain region introduced a new path to researchers, planners, engineer, etc. Various approaches and methods (physical models, rainfall-intensity model, statistical models, and probabilistic models) have been adopted as preparedness measures for destructive landslide phenomena. A useful model will be able to make reliable forecasts for environments other than those for which they were originally constructed (Kirkby 1996). Since the inception of RS and GIS, the preparation of landslide susceptibility map using various statistical techniques has become very much significant worldwide. In Darjeeling Himalaya, a large-scale landslide phenomenon sometimes plagued the socio-economic status of the common people. The spatial distribution of potential landslide-prone areas in Darjeeling Himalaya with the help of RS and GIS will help a lot for site-specific planning and development. In the present study, various geomorphic and hydrologic parameters such as slope, aspect, curvature, lineament, geology, geomorphology, soil, distance to drainage, drainage density, distance to lineament, lineament density, stream power index (SPI), topographic wetness index (TWI), land use and land cover (LULC), and rainfall were being incorporated for the assessment and prediction of landslide susceptible zones in Darjeeling Himalaya. In the present study, different models, i.e.frequency ratio (FR) model, modified information value (MIV) model, logistic regression (LR) model, artificial neural network (ANN) model, weighted overlay analysis (WOA) model, certainty factor (CF) model, analytical hierarchy process (AHP) model and fuzzy logic (FL) approach were adopted to assess and predict landslide susceptibility zones of Darjeeling Himalaya. To perform the models, previous knowledge on landslides causative factors (slope, elevation, soil, rainfall, etc.) and landslide inventory were incorporated to assign weights as well as to derive prioritized factor and class rating values corresponding to each class and factors. All the models forecasted potential landslide susceptible areas of Darjeeling

S. Mandal, S. Mondal, *Statistical Approaches for Landslide Susceptibility Assessment and Prediction*, https://doi.org/10.1007/978-3-319-93897-4_8

Himalaya with reasonable accuracy. Each and every model of the study used a particular type of algorithm to integrate data layers on GIS environment. The validation of all the models was done by developing ROC curve to assess the reliability based on the landslide data sets. It is to be inferred that GIS performed as a sophisticated tool for integrating all the required data layers responsible for landslides in Darjeeling Himalaya as well as developing a reliable outcome for planning and development.

8.2 Significance of Modelling in Geographical and Landslide Studies

The development of any discipline owes its origin to the development of its existing own philosophy. Philosophical-methodological models of geography are the basic elements of the scientific study of geography. It is associated with cognitive properties of geography which describe and determine its scientific character. According to Maik (2007), "The dispute over the right methodology of geography went on for the entire 20th century. Methodological discussions greatly contributed to the development of subject discipline of geography, especially in the second half of the 20th century". The philosophical studies considered by geographers to understand the cognitive nature of geography on various directions of philosophical assumptions. Various philosophical-methodological orientations of geography came into existence in the 1960s, i.e. empiricism, positivism, and neo-positivism, humanism, phenomenology, idealism, structuralism, Marxism and neo-Marxism, postmodernism, and feminism (Chojnicki 1985, 2004; Cloke et al. 1991; Peet 1998; Fotheringham 2006) in geographical studies. The empirical-scientific model of geography has formed under the influence of *empiricism* as a philosophical stream (Boven 1981). In geographical inquiry, the observation of facts and figures and formulation of scientific laws and theories started to appear with the application of statistical methods in making generalizations. According to Wilson (1972), "The inductive method involves theorizing from a mass of observations. In its most refined form, this is more or less coincident with statistical analysis". This adoption of quantitative, especially statistical, methods manifested itself mainly in spatial analysis concentrating largely on the examination of spatial structures and processes (Berry and Marble 1968; Chojnicki and Marble 1973). Systematic approach in geographical studies first propounded by Chorley (1962), Haggett (1965), and Chapman (1977) which demonstrates a structural framework of systematic models with reference to empirical-scientific model. Beven and Kirkby (1979), Montgomery and Dietrich (1994), and Dietrich et al. (1995, 2001) introduced so many process-based models in relation to landslides. In these approaches, the observation of geographical phenomena and its continuous changes are being observed which helps in experimental studies or in simulation modelling. The application of aerial photography, satellite-borne remote sensing, the Global Positioning System (GPS), and the Geographic

Information System to assess the state of geographical phenomena over the earth surface and its changes provided a strong empirical base of geographical knowledge (Ciołkosz 2007). In the present study, SRTM DEM, satellite imageries, Google earth images, and topographical maps were processed on GIS environment to prepare various data layers, i.e. slope, aspect, curvature, distance to lineament, lineament density, distance to drainage, drainage density, land use and land cover, stream power index, and topographic wetness index were developed. The identification of landslide scars and measurement of areal extent of landslide-prone areas were done with the help of GPS and extensive field work with necessary accessories. The acquired knowledge about various geomorphic and hydrologic attributes and their role in landslide activities helped to assign weights to each class and factor as well as to run various models and to prepare landslide susceptibility map of Darjeeling Himalaya. To perform the landslide probability models, the occurrences of landslide phenomena in the past become much more significant which is related with the identification of facts and figures and deductive approach. So, the geographical enquiry and development of data set on landslide occurrences and their locations helped to perform suitable statistical method and to assess accuracy result. The geographical knowledge of enquiry helped to determine the factors responsible for landslides and to prepare corresponding data layers efficiently. The geographical knowledge provided a concrete base to develop all the models, i.e. *frequency ratio model, modified information value model, logistic regression model, artificial neural network model, weighted overlay analysis model, certainty factor analysis, analytical hierarchy process*, and *fuzzy logic approach* and to prepare landslide susceptibility zonation map of Darjeeling Himalaya using GIS tool.

8.3 Comparison Between Various Statistical Models in Landslide Studies

The landslide susceptibility map of Darjeeling Himalaya was made applying *frequency ratio model, modified information value model, logistic regression model, artificial neural network model, weighted overlay analysis model, certainty factor model, analytical hierarchy process*, and *fuzzy logic approach.* The comparison between models was being made on the basis of ROC curve analysis. All the prepared landslide susceptibility maps of Darjeeling Himalaya were classified into five, i.e. very low, low, moderate, high, and very high. The produced maps were compared with the existing landslide locations in the Darjeeling Himalaya.

The prepared landslide susceptibility maps of Darjeeling Himalaya using *frequency ratio model* and *modified information value model* were grouped into very low, low, moderate, high, and very high landslide susceptibility zones. Lee and Sambath (2006) and Lee and Pradhan (2007) successfully used frequency ratio model to assess a correlation between the historical slide locations and various influencing factors under consideration. Sharma et al. (2009) concluded that the accuracy assessment of landslide susceptibility map confirmed the model with highest degree of accuracy. Akbar and Ha (2011) pointed out that Information Value Model is an integrated model for landslide susceptibility zonation mapping using a modified form of pixel, Global Positioning System (GPS), and GIS and RS. The derived AUC values of frequency ratio model and modified frequency ratio model experienced the prediction accuracy of 74.60% and 76.90%, respectively (Fig. 8.1). The study revealed that the modified information value model is suitable to prepare landslide susceptibility map of Darjeeling Himalaya in comparison to frequency ratio model.

The logistic regression is one of the most popular methods which showed the relationship between a dependent variable and several independent variables (Pradhan and Lee 2010). It is introduced by McFadden (1974) for the measurement of probability of any hazard or disaster using specific equation and landslide controlling variables. The values of LR are directly related to the probability of landslide occurrence. The probability of landslide occurrence was calculated using Eq. (8.1):

$$p = 1/\left(1 + e^{-z}\right) \quad \text{(Eq. 8.1)}$$

where p is the probability of landslide occurrence, e is the exponential, and z is the linear combination.

The area under curve (AUC) of the ROC curve was 0.715 or 71.5% (Fig. 8.2) which states that the logistic regression (LR) method used in the study showed a reasonable accuracy in predicting the landslide susceptibility of Darjeeling Himalaya. AUC value of ROC curve ranges from 0.5 to 1.0. AUC 1.0 means perfect prediction. To assess the relationship between causative factors and landslides, LR

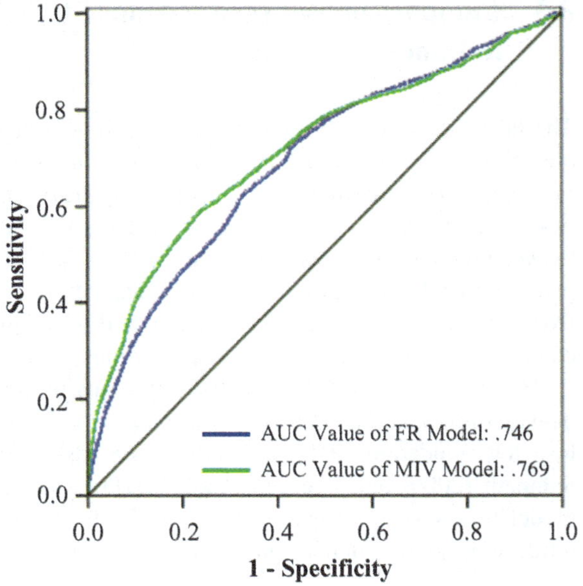

Fig. 8.1 ROC curve of frequency ratio model and modified information value model

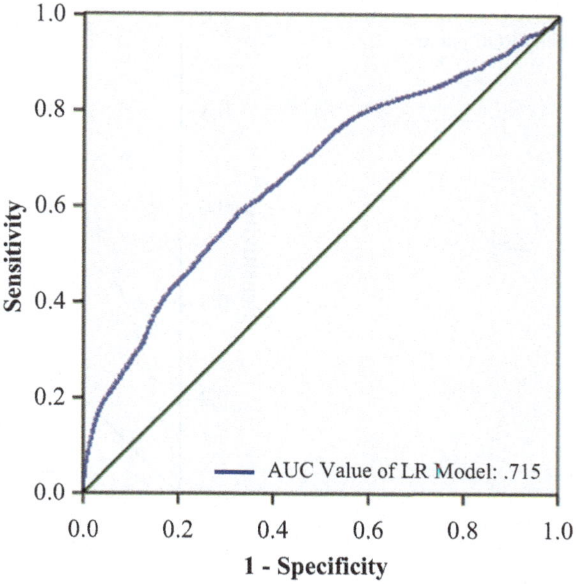

Fig. 8.2 ROC curve and accuracy result

co-efficient values were derived. Positive LR values were found in favour of soil, distance to lineament, lineament density, SPI, TWI, and rainfall which depicted a strong influence of all there attributes on slope instability.

Artificial Neural Network (ANN) is considered as machines which are designed in such a way that the human mind works to a specific task (Haykin 1999). Carrara et al. (1995) and Emami et al. (1998) applied artificial neural network model as a statistical approach in their studies. Lee et al. (2001) used ANN in landslide susceptibility assessment and prediction. The prepared landslide susceptibility zonation map of Darjeeling Himalaya using ANN model was validated with the help of derived ROC curve that showed the accuracy level of 81.5% (Fig. 8.3). The study showed that the application of ANN in landslide susceptibility zonation mapping can be of immense important in regional planning and development of Darjeeling Himalaya as the model introduced a high accuracy.

The prepared landslide susceptibility maps of Darjeeling Himalaya using *weighted overlay analysis, certainty factor model,* and *analytical hierarchy processes* were classified into very low, low, moderate, high, and very high landslide susceptibility zones. These three approaches are dealt with the estimation and assignment of weights to each class and factor for performing linear combination model on GIS environment and development of landslide susceptibility map. Shortliffe and Buchanan (1975) first used certainty factor approach in landslides assessment and prediction. Analytical hierarchy process was first introduced by Saaty (1980) which considered the logical assignment of values against each class and factors to derive class weight and factor weight. The accuracy level of weighted overlay analysis model, certainty factor approach, and analytical hierarchy process are 65.4%, 81.2%, and 67.5%, respectively (Fig. 8.4).

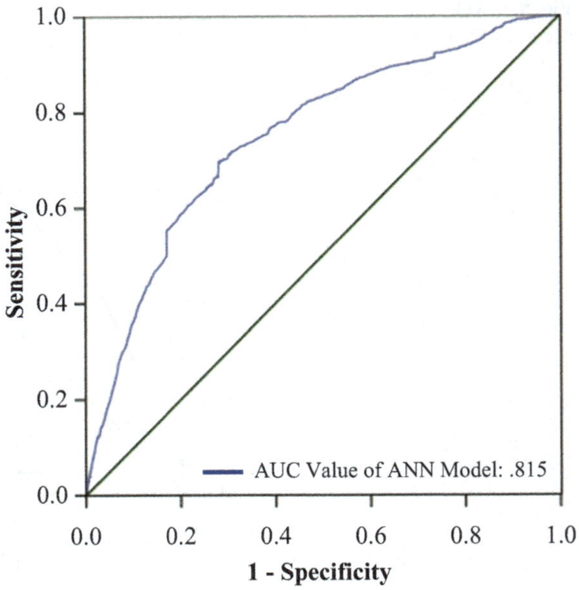

Fig. 8.3 Accuracy result using ROC curve

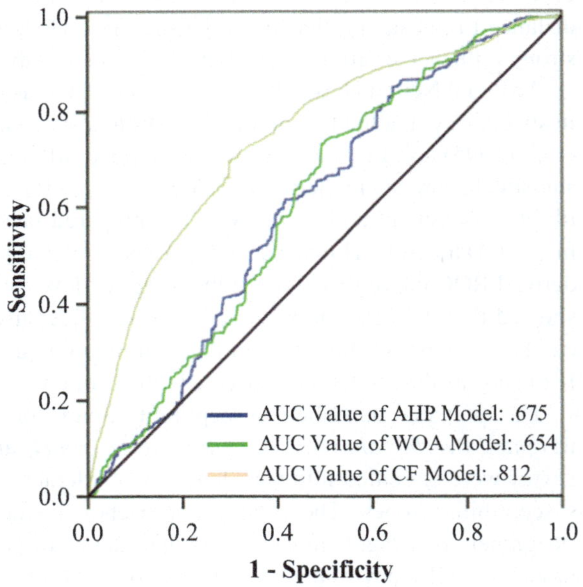

Fig. 8.4 ROC curve and validation result of WOA model, CF model, and AHP

Landslide susceptibility map of Darjeeling Himalaya was made based on both frequency ratio-based fuzzy gamma operator and cosine amplitude-based fuzzy gamma operator. Pradhan et al. (2009), Pradhan (2010a, 2011a), Kanungo et al. (2006), Bui et al. (2012d) applied fuzzy logic approach in landslide susceptibility studies using RS and GIS. Here, to derive fuzzy membership value frequency ratio

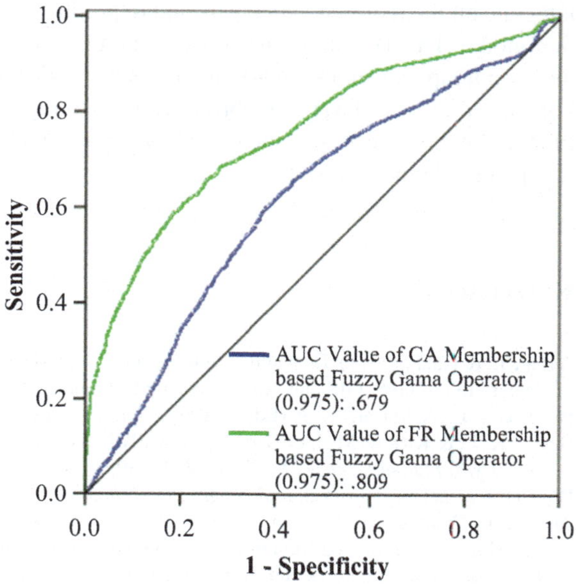

Fig. 8.5 ROC curve and models validation

method and cosine amplitude method were used. Frequency ratio-based fuzzy membership and landslide susceptibility zonation map of Darjeeling Himalaya showed the accuracy level of 80.9% and cosine amplitude-based fuzzy membership and landslide susceptibility depicted the accuracy level of 67.9%. Study depicted that the frequency ratio-based landslide susceptibility map having high accuracy level than cosine amplitude-based fuzzy membership function and developed landslide susceptibility (Fig. 8.5).

8.4 Conclusion

The estimated accuracy level of *frequency ratio model*, *modified information value model*, *logistic regression model*, *artificial neural network model*, *weighted overlay analysis model*, *certainty factor model*, *analytical hierarchy process model*, frequency ratio-based *fuzzy logic* and cosine amplitude-based *fuzzy logic* are 74.6%, 76.9%, 71.5%, 81.5%, 67.5%, 81.2%, 67.5%, 80.9%, and 67.9%, respectively. The study forecasted that artificial neural network model (ANN), certainty factor (CF) model, and frequency ratio-based fuzzy logic approach are most reliable statistical techniques in the assessment and prediction of landslide susceptibility in Darjeeling Himalaya. The application of GIS and introduction of various statistical models contributed a lot in the field of geographical enquiry. The validation of each model and prepared landslide susceptibility map with the help of observed facts and figures have become much more significant aspect in landslide sciences. The modelling in geographical studies with the help of GIS provided a scientific base to integrate all the required data layers and to produce a good result. The digital elevation model (DEM) is used and processed on GIS to generate slope, aspect, curvature, topographic wetness index, stream power index, etc. of a geographical region. For the fulfilment of various objectives in geographical research within a limited time frame, the role of statistical models is very much significant for the identification potential landslide sites.

References

Akbar, T., & Ha, S. (2011). Landslide hazard zoning along Himalaya Kaghan Valley of Pakistan by integration of GPS, GIS and remote sensing technology. *Landslides, 8*(4), 527–540.

Berry, B. J. L., & Marble, D. F. (Eds.). (1968). *Spatial analysis*. Inglewood Cliffs: Prentice Hall.

Beven, K. J., & Kirkby, M. J. (1979). A physically based contributing area model of basin hydrology. *Hydrological Sciences Bulletin, 24*, 43–69.

Boven, M. (1981). *Empiricism and geographical thought*. Cambridge: University Press.

Bui, D. T., Pradhan, B., Lofman, O., Revhaug, I., & Dick, O. B. (2012). Spatial prediction of landslide hazards in Hoa Binh province (Vietnam): a comparative assessment of the efficacy of evidential belief functions and fuzzy logic models. *Catena, 96*, 28–40.

Carrara, A., Cardinali, M., Guazzetti, F., & Reichenbach, P. (1995). GIS techniques in mapping landslide hazard. In A. Carrara & F. Guzzetti (Eds.), *Geographical information systems in assessing natural hazards* (pp. 135–175). The Netherlands: Kluwer Academic Publishers.

Chapman, G. P. (1977). *Human and environmental systems. A geographer's appraisal*. London: Academic Press.

Chojnicki, Z. (1985). Orientacje filozoficzno-metodologiczne geografii – ich koncepcje i model (Philosophical-methodological orientations in geography: conceptions and models). *Przegląd Geograficzny, 57*(3), 255–281.

Chojnicki, Z. (2004). Podstawy filozoficzne geografii – jakiej filozofii potrzebuje geografia? (Philosophical foundations of geography: What philosophy does geography need?). In Z. Chojnicki (Ed.), *Geografia wobec problemów teraźniejszości i przyszłości* (pp. 191–207). Poznań: Bogucki Wyd Naukowe.

Chojnicki, Z., & Marble D. F. (Eds) (1973) *Perspectives on spatial analysis*. Geographia Polonica 25.

Chorley, R. (1962). *Geomorphology and general systems theory*. Geological Survey Professional Paper, 500 B.

Ciołkosz, A. (2007). Nowe narzędzia i metody badawcze w geografii oraz ich rola w rejestracji i wyjaśnianiu zjawisk i zmian w skali globalnej (New research tools and methods in geography and their role in the registration and explanation of phenomena and changes at a global scale). In W. Maik, K. Rembowska, & A. Suliborski (Eds.), *Geografia a przemiany współczesnego świata* (pp. 103–118). Bydgoszcz: Wyd. Uczelniane WSG.

Cloke, P., Philo, C., & Sadler, D. (1991). *Approaching human geography*. London: Paul Chapman Publishing.

Dietrich, W. E., Reiss, R., Hsu, M. L., & Montgomery, D. R. (1995). A process-based model for colluviums soil depth and shallow landsliding using digital elevation model data. *Hydrological Processes, 9*, 383–400.

Dietrich, W. E., Bellugi, D., & Real de Asua, R. (2001). Validation of shallow landslide model, SHALSTAB, for forest management. In M. S. Wigmosta & S. J. Burges (Eds.), *Land use and watersheds: human influence on hydrology and geomorphology in urban and forest areas, Water science and application monograph* (Vol. 2, pp. 195–227). Washington, DC: American Geophysical Union.

Emami, S. M. R., Iwao, Y., & Harada, T. (1998) Performance of Artificial Network system in prediction issues of earthquake engineering. *Proceedings of 8th International IAEG Congress* (pp. 733–738).

Fotheringham, A. S. (2006). Quantification, evidence and positivism. In S. Aitken & G. Valentine (Eds.), *Approaches to human geography* (pp. 237–250). London: Sage.

Haggett, P. (1965). *Locational analysis in human geography*. London: Arnold.

Haykin, S. (1999). *Neural networks. A comprehensive foundation* (2nd ed.p. 696). Englewood Cliffs, NJ: Prentice Hall.

Kanungo, D. P., Arora, M. K., Sarkar, S., & Gupta, R. P. (2006). A comparative study of conventional, ANN black box, fuzzy and combined neural and fuzzy weighting procedures for landslide susceptibility zonation in Darjeeling Himalayas. *Engineering Geology, 85*, 347–366.

Kirkby, M. J. (1996). A role for theoretical models in Geomorphology? The Scientific Nature of Geomorphology. In B. L. Rhoads & C. E. Thorn (Eds.), *Proceedings of the 27th Binghamton Symposium in Geomorphology held 27–29 September 1996*. Chichester: John Wiley & Sons Ltd.

Lee, S., & Pradhan, B. (2007). Landslide hazard mapping at Selangor, Malaysia using frequencyratio and logistic regression models. *Landslides, 4*(1), 33–41.

Lee, S., Ryu, J., Min, K., & Won, J. (2001). *Proceedings of the Geoscience and Remote Sensing Symposium, IGARSS '01* (vol. 5, pp. 2364–2366). IEEE 2001 International.

Lee, S. & Sambath, T. (2006). Landslide Susceptibility mapping in the Damrei Romel area, Cambodia using frequency ratio and logistic regression models. *Environmental Geology, 50*, 847–856.

Maik, W. (2007). Geografia a współczesność w świetle tradycji myśli geograficznej (Geography and modernity in the light of the tradition of geographical thought). In W. Maik, K. Rembowska, & A. Suliborski (Eds.), *Geografia a przemiany współczesnego świata* (pp. 25–37). Bydgoszcz: Wyd. Uczelniane WSG.

McFadden, D. (1974). Conditional logit analysis of qualitative choice analysis. In P. Zarembka (Ed.), *Frontiers in econometrics* (pp. 105–142). New York: Academic Press.

Montgomery, D. R., & Dietrich, W. E. (1994). A physically based model for the topographic control on shallow landsliding. *Water Resources Research, 30*(4), 1153–1171.

Peet, R. (1998). *Modern geographical thought*. Oxford: Blackwell Publishing.

Pradhan, B. (2010). Application of an advanced fuzzy logic model for landslide susceptibility analysis. *International Journal of Computational Intelligence Systems, 3*(3), 370–381.

Pradhan, B., & Lee, S. (2010). Regional landslide susceptibility analysis using back propagation neural network model at Cameron Highland, Malaysia. *Landslides, 7*, 13–30.

Pradhan, B., Lee, S., & Buchroithner, M. F. (2009). Use of geospatial data for the development of fuzzy algebraic operators to landslide hazard mapping: A case study in Malaysia. *Applied Geomatics, 1*, 3–15.

Saaty, T. L. (1980). *The analytical hierarchy process* (p. 350). New York: McGraw Hill.

Sharma, L., Patel, N., Ghosh, M., & Debnath, P. (2009). Geographical information system based landslide probabilistic model with trivariate approach - a case study in Sikkim Himalaya. *Eighteenth United Nations Regional Cartographic Conference for Asia and the Pacific*, UN, Bankok, Economic and Social Council.

Shortliffe, E. H., & Buchanan, B. G. (1975). A model of inexact reasoning in medicine. *Mathematical Biosciences, 23*(3), 351–379.

Wilson, A. G. (1972). Theoretical geography: Some speculations. *Transactions of the Institute of British Geographers*, (57), 32–44.

[illegible]

Prakash [illegible]

Pradhan [illegible]

Shah [illegible]

Sharma [illegible]

Sinha [illegible]

Singh [illegible]

Index

S. Mandal, S. Mondal, *Statistical Approaches for Landslide Susceptibility Assessment and Prediction*, https://doi.org/10.1007/978-3-319-93897-4

The manufacturer's authorised representative in the EU is Springer Nature Customer Service Centre GmbH, Europaplatz 3, 69115 Heidelberg, Germany. If you have any concerns regarding our products, please contact ProductSafety@springernature.com

Printed and bound by CPI Group (UK) Ltd, Croydon, CR0 4YY
30/09/2026
02225662-0001